# SpringerBriefs in Electrical and Computer Engineering

## Control, Automation and Robotics

*Series Editors*

Tamer Başar
Antonio Bicchi
Miroslav Krstic

For further volumes:
http://www.springer.com/series/10198

Adnan Tahirovic · Gianantonio Magnani

# Passivity-Based Model Predictive Control for Mobile Vehicle Motion Planning

 Springer

Adnan Tahirovic
Faculty of Electrical Engineering
University of Sarajevo
Sarajevo
Bosnia and Herzegovina

Gianantonio Magnani
Dipartimento di Elettronica
Politecnico di Milano
Milan
Italy

ISSN 2192-6786
ISBN 978-1-4471-5048-0
DOI 10.1007/978-1-4471-5049-7
Springer London Heidelberg New York Dordrecht

ISSN 2192-6794 (electronic)
ISBN 978-1-4471-5049-7 (eBook)

Library of Congress Control Number: 2013934010

Printed on acid-free paper

Springer is part of Springer Science+Business Media (www.springer.com)

*To Enver, Djula and Adaleta*

AD

*To Giuliana, Carlo and Silvia*

MG

# Acknowledgments

We would like to thank Karl Iagnemma (Massachusetts Institute of Technology), Luca Bascetta (Politecnico di Milano), Paolo Rocco (Politecnico di Milano), and Yoshiaki Kuwata (NASA—Jet Propulsion Laboratory) for their valuable help and comments and the discussions on the topics related to this field, which helped enrich and complete the content of this work. Our sincere thanks go to Dinko Osmankovic and Anel Husakovic (Faculty of Electrical Engineering Sarajevo) for helping us with the ACADO simulations related to the last section of the Brief. We feel deeply grateful to our families and friends for their unconditional support and inexhaustible motivation. Finally, we would like to express our gratitude to Oliver Jackson from Springer-Verlag and the Series Editor for their valuable support.

# Contents

# Abstract

An autonomous mobile vehicle able to traverse a wide range of poor natural terrains is an indubitable useful concept that can be utilized in different kinds of fields. A variety of needs of planetary explorations, rescue missions in hazard areas, humanitarian demining as well as agriculture applications have recently triggered a lot of research works aiming at developing sufficiently reliable motion and navigation planning approaches in such environments. This work presents such a concept to guide the vehicle to reach a goal position regardless the terrain shapes.

The presented navigation planning approach is based on the model predictive control paradigm (MPC). The MPC-like approaches allow for taking into account a variety of different constraints, such as guaranteeing stability, avoiding obstacles, and preventing the vehicle from sideslip and rollover. An additional feature of an MPC approach is that it continuously repeats the optimization during the task execution allowing for new local sensor measurements to be taken into account. Such a policy is used for continuous finding collision-free paths and to guarantee the safe task execution. Additionally, it inherently provides a certain level of robustness to an MPC generated path comparing to the approaches where the complete path is being found prior the task execution.

In order to adopt an MPC-like approach for the purpose of mobile vehicle navigation, we use energy shaping technique to include the terrain map and the goal position into the system model. The passivity-based control theory is then used to obtain a stable MPC framework (PB/MPC) guaranteeing task completion, which means the vehicle is being capable to reach the goal position.

The straightforward procedure for finding feasible control actions, regardless the complexity of the vehicle model, makes this approach a good tool to be used in outdoor environments. Namely, using a precise complex model that reflects the vehicle behavior on rough terrains, during the planning stage, provides a safer planner which generates trajectories that can be easily tracked by the vehicle during the execution stage. The problem of using a simplified model to generate trajectories during the planning stage is certainly an issue of the navigation planning for complex vehicles and environments.

# Chapter 1
# Introduction

## 1.1 Motivation

The popularity of the research on unmanned ground vehicles has increased recently due to a variety of operations and environments. Planetary explorations, search and rescue missions in hazard areas, surveillance, humanitarian de-mining, as well as agriculture applications such as pruning vine and fruit trees, represent possible fields of using autonomous vehicles in natural environments. Planetary exploration allows for understanding the planet surface geology, its present and past climate conditions, and for discovering potential signs of other lives. Rescue missions in dangerous environments and surveillance with reduced human operations and interventions have also become interesting areas for unmanned vehicles. With the special interest within the military industry, these areas are growing rapidly and motivate new research in autonomous vehicle motion planning in difficult environments. The use of the autonomous vehicles in a de-mining process decreases the danger and the cost of manual mine detection. Both humanitarian and economic motivations to use such vehicles to this purpose are obvious. Finally, agriculture applications have recently recognized the potential to use the fully autonomous vehicles in agricultural operations reducing the total cost of the final product. Some of the respective vehicles used for different applications are shown in Figs. 1.1, 1.2, 1.3, and 1.4.

## 1.2 Motion Planning Literature

During the past 20 years, motion planning has become one of the most active research areas in robotics. A nice overview of motion planning has been presented in [1, 2]. The main focus of the early research stage was finding collision-free paths. In [3–5], the potential field approach for real-time obstacle avoidance was introduced and discussed both for manipulators and mobile ground robots. In order to define the final potential field, this technique defines goals and obstacles as attractive and repulsive

A. Tahirovic and G. Magnani, *Passivity-Based Model Predictive Control for Mobile Vehicle Motion Planning*, SpringerBriefs in Control, Automation and Robotics, DOI: 10.1007/978-1-4471-5049-7_1, © The Author(s) 2013

**Fig. 1.1**   NASA/JPL mars explorations rover

**Fig. 1.2**   Different NASA/JPL explorations rovers

forces, respectively, in order to navigate the vehicle toward the decrease of the corresponding potential function. Artificial potential fields have been used in variety of applications including mobile vehicles in natural terrains [1, 4, 6–8]. Ge and Cui dealt with the problem of moving obstacles using the potential field method [9]. The main drawback of the approach based on artificial potential fields is the existence of local minima that may trap the robot in an undesired location. An improvement of computational efficiency of a method that deals with local minima was analyzed in [10], where a parallel computational scheme was illustrated. The local minima problem arisen from potential field method is still attractive subject for many researchers. Some of the methods carefully compute potential functions to contain only a unique

**Fig. 1.3** Surveillance vehicles: 1. Seekur UGV from MobileRobots Inc, 2. μTrooper from Thales, and 3. MDARS from general dynamics

**Fig. 1.4** Humanitarian demining vehicles: 1. Buggy-robot from Hirose and Fukushima laboratory, 2. Humi from TU Wien, and 3. Gryphon-III with field arm from Hirose and Fukushima laboratory

global minimum at the goal position [11–13]. The approach of Laplace's equation [11] is considered to be computationally expensive. The concept of navigation functions [12, 13] was successfully used to make a potential field in Cartesian space to navigate mobile robots. How to efficiently recompute a navigation function in a dynamical environment is shown [14]. A new class of navigation functions that are appropriate for nonholonomic motion planning were presented in [15]. High speed navigation of mobile vehicles on uneven terrains was introduced in [16], where an artificial field is computed using rollover and sideslip constraints, hazard locations, as well as desired velocities.

In [17], authors have found the way to compute the fastest dynamically admissible speed along a given path on a three-dimensional terrain considering dynamics such as slip, rollover, and ballistic motion. Finding a time optimal path using the result of this work is presented in [18], while static and moving obstacles are considered in [19]. A drawback of these methods is computational inefficiency for real-time applications. Some different techniques for collision avoidance taking into account the vehicle velocity constraints have also been presented in [20] and [21].

The research on motion planning evolved by adding the capability of taking into account the vehicle motion dynamics constraints within the well-known dynamic window approach (DWA) [22, 23]. The DWA selects translational and rotational velocities by maximizing an objective function based on the vehicle heading to the goal position, distance to the closest obstacle, and velocity of the vehicle. The optimization is performed using arcs considering only reachable and safe velocities. This subject was extended to the high-speed navigation of a mobile robot in [14] by the global DWA, as the generalization of the DWA. A combination of the DWA with other methods yielded some improvements in long-term real-world applications [24]. Dubowsky and Iagnemma extended the DWA to rough terrains using the vehicle curvature-velocity space bounded by hazards as well as steering limits, wheel-terrain interaction, rollover and sideslip constraints. In this space the stability constraints of the vehicle, for instance, expressed by limit values of the roll-over and side slip indexes, can be easily described. The given algorithm was also suitable for high speed vehicles and appropriate for real-time implementation [16, 25, 26]. A convergent DWA was obtained for the unicycle mobile vehicle [27] exploiting the model predictive control combined with the direct Lyapunov function approach (MPC/CLF) [27, 28].

Sequences of motion primitives have been used to cover local planning search space since [29]. More recent works are given in [30] and [31], where the inverse trajectory generation was used to navigate UAV and UGV, respectively. The importance of separation in a local planning search space is discussed in [32] and it was shown that the mutual separation of a set of paths is related to the relative completeness of the motions set. The planning approach proposed in [33] generate path sets to navigate an UGV. The planner considers global guidance, satisfies environmental constraints, and guarantees dynamic feasibility by the use of a model-predictive trajectory generator. A grid-based planning approach which takes into account the vehicle differential constraints is introduced in [34]. This planner uses the vehicle model to generate the state lattices assuring the feasible paths along the cost map edges. The heuristic cost estimate, which represents cost-to-go for each node of the grid needed by the $A^*$ algorithm [35], is taken from a priori calculated heuristic look-up table (HLUT) [36, 37], which is based on the path length, speeding up the algorithm. In [34], the cost edges are calculated by solving two boundary value problem where the control action is parameterized converting the problem into a nonlinear programming one. The cost function represented the minimum slope-dwell performance index, selecting less difficult paths between the nodes (lattice states) that are considered in the overall optimization. Including the vehicle model into the motion planning stage provides a planner which generates trajectories that can be easily followed by a mobile robot. This especially comes to the fore when a vehicle moves with high speed and operates on rough terrains. Using a simpler planner that does not take into account the mobile vehicle model might cause a fatal error due to the difference between the planned and executed trajectories. For this reason, the gradient-based algorithms such as the navigation function or a variant of the $D^*$ [38, 39], in our case are not considered being an acceptable solution.

The sample-based technique for robot motion planning was introduced in [40]. The first sample-based motion planners were not computationally efficient for certain environments. In [41–43] the probabilistic roadmap method (PRM) was developed for path planning in configuration spaces with many degrees of freedom. A comprehensive overview and discussion about PRM is given in [44] and [45].

PRM method has proven to work well in static well-known environments and are considered computationally efficient for car-like vehicles [46]. However, PRM may not be suitable for planning in a dynamic environment, especially because it does not take into account the vehicle dynamics and might result in very sharp turning points. In [47] the authors introduced quasi-PRM and lattice roadmap (LRM) algorithms. LRM was extended in [34] to allow the state lattice to represent the differential constraints of the mobile vehicle. Rapidly exploring random trees (RRT) is a type of probabilistic planners originally developed to cope with differential constraints [48–50]. A significant feature of the RRT-like algorithms is that the resulting trajectories are executable by the underlying dynamical system. The RRT algorithm has been proven probabilistically complete [50], meaning that the probability of finding a solution feasible path converges to one if such a path exists. An improvement of the RRT algorithm was proposed in [51], where the obtained exponential convergence speed yielded a good performance. Several variants of the roughness-based RRTs are illustrated in [52–54], while some recent results on the RRT-like planners have been introduced in [55] and [56].

The RHC/CLF (Receding Horizon Control/Control Lyapunov Function) scheme developed in [57] used the concept of control Lyapunov function to obtain the stability of RHC scheme. The authors presented the generalization of the RHC/CLF scheme demonstrating its relation to the optimal controller. In [27], the authors have implemented the same scheme (MPC/CLF, Model Predictive Control/Control Lyapunov Function) for the navigation planning of a unicycle mobile vehicle. The approach developed and proposed in [58] used a passivity-based constraint to obtain an MPC scheme with guaranteed closed loop stability for nonlinear systems. Inspired by this control concept, a framework for mobile robot motion planning using the PB/MPC is presented both for flat and rough terrains in [59–61]. The PB/MPC motion planning framework will be the focus of this work.

## 1.3  Passivity-Based Control Overview

The passivity theory has been used in the stability analysis of nonlinear electrical circuit networks, see e.g., [62]. The state-space representation with its storage and dissipation energy described by Lyapunov functions was used in [63] and [64]. Pioneer works on the connection among stabilization, Lyapunov function existence, and optimality within the passivity theory have appeared in [65] and [66]. A basic stabilization property of interconnected passive systems has been introduced in [67] and [68]. These papers have triggered a large number of research works in the passivity-based control field. The idea of designing a controller that stabilizes the system using the framework of total energy shaping combined with the passivity-based approach was given in [69].

In order to introduce the passivity-based control concept, let us consider the system

$$\dot{x} = f(x, u) \tag{1.1}$$
$$y = h(x),$$

where the origin $x = 0$ is an open-loop equilibrium point, $h(0) = 0$, $f$ is locally Lipschitz in $(x, u)$ and $h$ is continuous in $x$, for all $x \in R^n$ and $u \in R^m$. If a continuously differentiable positive semidefinite function $V(x)$ exists representing the storage function, so that

$$u^T y \geq \dot{V} = \frac{\partial V}{\partial x} f(x, u), \ \forall (x, u) \in R^n \times R^m, \tag{1.2}$$

then the system (1.1) is passive.

A basic idea on the passivity-based control is expressed by the following theorem, recalled without proof (see e.g., [70]).

**Theorem 1** *If system (1.1) is passive with a radially unbounded positive definite storage function, and zero-state observable, then the origin $x = 0$ can be globally stabilized by $u = -\phi(y)$, where $\phi$ is any locally Lipschitz function such that $\phi(0) = 0$ and $y^T \phi(y) > 0$ for all $y \neq 0$.*

The following system (1.3) is a special case of (1.1). It can be transformed into a passive system choosing the output (1.4), where the storage function $V(x)$ is a radially unbounded, positive definite, continuously differentiable function satisfying (1.5).

$$\dot{x} = f(x) + G(x)u \tag{1.3}$$

$$y = h(x) = \left[\frac{\partial V}{\partial x} G(x)\right]^T \tag{1.4}$$

$$\frac{\partial V}{\partial x} f(x) \leq 0, \forall x \tag{1.5}$$

## 1.4 Passivity-Based Model Predictive Control

A variant of the passivity-based nonlinear model predictive control scheme for the system (1.1) is given by the setup

$$\inf_{u(\cdot)} \int_{t_0}^{t_0+T} (q'(x) + u^T u)dt \tag{1.6}$$

$$\dot{x} = f(x) + g(x)u \tag{1.7}$$
$$y = h(x) \tag{1.8}$$
$$u^T(t)y(t) < -y^T(t)\phi(y), \tag{1.9}$$

where $q'(x)$ is a state cost function. In the PB/MPC optimization setup proposed in [58], $\phi(y) = y$. The choice of the passivity-based constraint (1.9) allows for the system stabilization with the feedback $u = -y$ if system (1.7) is passive and zero-state detectable. It will be shown that the stabilization can be also achieved using any function $\phi(y)$ defined in Theorem 1 and using the general feedback law $u = -\phi(y)$ if system (1.7) is passive and zero-state observable.

Differently than the MPC/CLF optimization setup, where the stability is achieved by enforcing a decrease of the Lyapunov function along the solution trajectory, the PB/MPC optimizations setup obtains the stability by using the passivity-based constraint.

As in case of the MPC/CLF, it has been shown in [58] that the proposed PB/MPC optimization setup also relates to the optimal control. This relationship is given by the following corollary recalled here without the proof.

**Corollary** *The optimal performance of the infinite horizon optimal control problem is recovered by the passivity-based model predictive control scheme (1.6–1.9 ), if $y = \frac{1}{2}g^T(x)\frac{\partial V^*}{\partial x}$ and $T \to 0$, where $V^*$ is the value function of the infinite horizon optimal control problem.*

## 1.5 Scope of the Work

In this work, we present the passivity-based model predictive control (PB/MPC) adopted from the control theory to the mobile vehicle navigation framework for both indoor and outdoor environments. The PB/MPC navigation approach stabilizes the goal position using both energy shaping technique by the use of navigation function and the passivity control concept. Stabilization of the goal position guarantees the task completion which means that the vehicle is able to reach the goal for a given environment. Unlike an MPC framework which is stabilized using the control Lypunov function (CLF) as a terminal constraint, and in which the feasible control actions have to be found for each optimization cycle in order to satisfy the stability constraint, the feasible control actions obtained by the PB/MPC is a direct consequence of the passivity-based approach. This means that the feasible control set obtained by the PB/MPC, over which the MPC optimization is performed, does not require any additional effort to be found. Such a property makes the PB/MPC navigation approach general and suitable to be used for a wide range of vehicles. For this reason, the PB/MPC can also be easily adapted to flat as well as rough terrains, where a truly complex vehicle that comprises the terrain model might be used to predict trajectories within the MPC optimization horizon. Obtaining feasible trajectories for a complex vehicle model, which can be safely tracked by the vehicle, is certainly a critical issue present in any motion planning approach in complex environments.

Chapter 2 presents the PB/MPC motion planning framework. Chapter 3 contains different examples and illustrates the usage of the planner. In Chap. 4, we analyze some limitations for the case of rough terrains and present a possible real-time implementation of an MPC-like motion planner. Chapter 5 outlines the presented work.

# References

1. J. Latombe, *Robot Motion Planning* (Kluwer, Boston, 1991)
2. S.M. LaValle, *Planning Algorithms* (Cambridge University Press, Cambridge, 2006)
3. O. Khatib, Real-time obstacle avoidance for manipulators and mobile robots. Int. J. Rob. Res. **5**(1), 90–98 (1986)
4. H. Haddad, M. Khatib, S. Lacroix, R. Chatila, Reactive navigation in outdoor environments using potential fields. in *Proceedings of the IEEE International Conference on Robotics and Automation*, pp. 1232–1237 (1998)
5. Y. Koren, J. Borenstein, Potential field methods and their inherent limitations for mobile robot navigation. in *Proceedings of the IEEE International Conference on Robotics and Automation*, pp. 1398–1404 (1991)
6. B. Chanclou, A. Luciani, Global and local path planning in natural environment by physical modeling. in *Proceedings of the 2010 IEEE/RSJ International Conference on Intelligent Robots and Systems*, pp. 1118–1125 (1996)
7. S. Sekhavat, M. Chyba, Nonholonomic deformation of a potential field for motion planning. in *Proceedings of the IEEE International Conference on Robotics and Automation*, pp. 817–822 (1999)
8. K.P. Valavanis, T. Hebert, R. Kolluru, N. Tsourveloudis, Mobile robot navigation in 2-d dynamic environments using an electrostatic potential field. IEEE Trans. Syst. Man Cybern. Part A: Syst. Hum. **30**(2), 187–196 (2000)
9. S.S. Ge, Y.J. Cui, Dynamic motion planning for mobile robots using potential field method. Auton. Robots **13**(3), 207–222 (2002)
10. S. Caselli, M. Reggiani, R. Sbravati, Parallel path planning with multiple evasion strategies. in *Proceedings of the IEEE International Conference on Robotics and Automation*, vol. 1, pp. 260–266 (2002)
11. C.I. Connolly, J.B. Burns, R. Weiss, Path planning using Laplace's equation. in *Proceedings of the IEEE International Conference on Robotics and Automation*, pp. 2102–2106 (1990)
12. E. Rimon, Exact robot navigation using artificial potential functions, Ph.D. thesis, 1990
13. E. Rimon, D.E. Koditschek, Exact robot navigation using artificial potential fields. IEEE Trans. Rob Autom. **8**, 501–518 (1992)
14. O. Brock, O. Khatib, High-speed navigation using the global dynamic window approach. in *Proceedings of the IEEE International Conference on Robotics and Automation*, vol. 1, pp. 341–346 (1999)
15. H. Tanner, S. Loizou, K. Kyriakopoulos, Nonholonomic navigation and control of cooperating mobile manipulators. IEEE Trans. Robot. Autom. **19**(1), 53–64 (2003)
16. S. Shimoda, Y. Kuroda, K. Iagnemma, High-speed navigation of unmanned ground vehicles on uneven terrain using potential fields. Robotica **25**(4), 409–424 (2007)
17. Z. Shiller, J. Chen, Optimal motion planning of autonomous vehicles in three dimensional terrains. in *Proceedings of the IEEE International Conference on Robotics and Automation*, pp. 198–203 (1990)
18. Z. Shiller, Y.-R. Gwo, Dynamic motion planning of autonomous vehicles. IEEE Trans. Robot. Autom. **7**(2), 241–249 (1991)
19. P. Fiorini, Z. Shiller, Motion planning in dynamic environments using velocity obstacles. Int. J. Robot. Res. **17**, 760–772 (1998)
20. J. Borenstein, Y. Koren, The vector field histogram—fast obstacle avoidance for mobile robots. IEEE J. Robot. Autom. **7**(3), 278–288 (1991)
21. J. Minguez, L. Montano, Nearness diagram (nd) navigation: collision avoidance in troublesome scenarios. IEEE Trans. Robot. Autom. **20**(1), 45–59 (2004)
22. R. Simmons, The curvature-velocity method for local obstacle avoidance, in *Proceedings of the IEEE International Conference on Robotics and Automation*, pp. 3375–3382 (1996)
23. D. Fox, W. Burgard, S. Thrun, The dynamic window approach to collision avoidance. IEEE Robot. Autom. Mag. **4**, 23–33 (1997)

24. R. Philippsen, R. Siegwart, Smooth and efficient obstacle avoidance for a tour guide robot. in *Proceedings of the IEEE International Conference on Robotics and Automation*, vol. 1, pp. 446–451 (2003)
25. M. Spenko, Y. Kuroda, S. Dubowsky, K. Iagnemma, Hazard avoidance for high speed unmanned ground vehicles in rough terrain. J. Field Robot. **23**(5), 311–331 (2006)
26. M. Spenko, Hazard avoidance for high speed rough terrain unmanned ground vehicles, Ph.D. thesis, Massachusetts Institute of Technology, MA, 2005
27. P. Oegren, N.E. Leonard, A convergent dynamic window approach to obstacle avoidance. IEEE Trans. Robot. **21**(2), 188–195 (2005)
28. P. Oegren, N.E. Leonard, A provably convergent dynamic window approach to obstacle avoidance. in *Proceedings of IFAC World Congress*, pp. 595–600 (2001)
29. L.E. Dubins, On curves of minimal length with a constraint on average curvature, and with prescribed initial and terminal positions and tangents. Am. J. Math. **79**(3), 497–516 (1957)
30. E. Frazzoli, M.A. Dahleh, E. Feron, Real-time motion planning for agile autonomous vehicles. J. Guidance Control Dyn. **1**(25), 116–129 (2002)
31. T.M. Howard, A. Kelly, Optimal rough terrain trajectory generation for wheeled mobile robots. Int. J. Rob. Res. **26**(2), 141–166 (2007)
32. C.J. Green, A. Kelly, Toward optimal sampling in the space of paths. in *Proceedings of the International Symposium of Robotics Research*, pp. 171–180 (2007)
33. T.M. Howard, C.J. Green, A. Kelly, D. Ferguson, State space sampling of feasible motions for high-performance mobile robot navigation in complex environments. J. Field Robot. **25**(10), 325–345 (2008)
34. M. Pivtoraiko, R.A. Knepper, A. Kelly, Differentially constrained mobile robot motion planning in state lattices. J. Field Robot. **26**(3), 308–333 (2009)
35. P. Hart, N. Nilsson, B. Raphael, A formal basis for the heuristic determination of minimum cost paths. IEEE Trans. Syst. Sci. Cybern. **4**(2), 100–107 (1968)
36. M. Pivtoraiko, A. Kelly, Efficient constrained path planning via search in state lattices. in *Proceedings of the 8th International Symposium on Artificial Intelligence, Robotics and Automation in Space*, vol. 8, Sept 2005
37. R. A. Knepper, A. Kelly, High performance state lattice planning using heuristic look-up tables. in *Proceedings of 2006 IEEE/RSJ International Conference on Intelligent Robots and Systems*, pp. 3375–3380, Oct 2006
38. A. Stentz, The focussed D* algorithm for real-time replanning. in *Proceedings of the International Joint Conference on Artificial Intelligence*, pp. 1652–1659 (1995)
39. S. Koenig, M. Likhachev, Fast replanning for navigation in unknown terrain. IEEE Trans. Robot. **21**(3), 354–363 (2005)
40. J. Barraquand, J. Latombe, Motion planning: a distributed representation approach. Int. J. Robot. Res. **10**(6), 628–649 (1991)
41. L.E. Kavraki, Random networks in configuration space for fast path planning, Ph.D. thesis, 1995
42. M.H. Overmars, P. Švestka, A probabilistic learning approach to motion planning. in *WAFR: Proceedings of the Workshop on Algorithmic Foundations of Robotics* (1995)
43. L. Kavraki, P. Svestka, J. Latombe, M. Overmars, Probabilistic roadmaps for fast path planning in high dimensional configuration spaces. IEEE Trans. Robot. Autom. **12**, 566–580 (1996)
44. H. Choset, K.M. Lynch, S. Hutchinson, G. Kantor, W. Burgard, L. Kavraki, S. Thrun, *Principles of Robot Motion: Theory, Algorithms, and Implementations* (MIT Press, Cambridge, 2005)
45. D. Hsu, J.-C. Latombe, H. Kurniawati, On the probabilistic foundations of probabilistic roadmap planning. Int. J. Rob. Res. **25**(7), 627–643 (2006)
46. G. Song, N.M. Amato, Randomized motion planning for car-like robots with C-PRM. in *Proceedings of the 2001 IEEE/RSJ International Conference on Intelligent Robots and Systems*, pp. 37–42 (2001)
47. M.S. Branicky, S.M. Lavalle, K. Olson, L. Yang, Quasi-randomized path planning. in *Proceedings of the IEEE International Conference on Robotics and Automation*, pp. 1481–1487 (2001)

48. J. Kuffner, S.M. Lavalle, Randomized kinodynamic planning. in *Proceedings of the IEEE International Conference on Robotics and Automation*, pp. 473–479 (1999)

49. S.M. Lavalle, J. Kuffner, Rapidly-exploring random trees: progress and prospects. in *Algorithmic and Computational Robotics: New Directions* (AK Petetrs Ltd., Welleslry, 2001), pp. 293–308

50. S.M. Lavalle, J. Kuffner, RRT-connect: an efficient approach to single-query path planning. in *Proceedings of the IEEE International Conference on Robotics and Automation*, pp. 995–1001 (2000)

51. D. Hsu, R. Kindel, J.-C. Latombe, S. Rock, Randomized kinodynamic motion planning with moving obstacles. in *Algorithmic and Computational Robotics: New Directions*, pp. 247–264 (2001)

52. D.J. Spero, R.A. Jarvis, Path planning for a mobile robot in a rough terrain environment. in *Third International Workshop on Robot Motion and Control*, pp. 9–11 (2002)

53. M. Kobilarov, G.S. Sukhatme, Time optimal path planning on outdoor terrain for mobile robots under dynamic constraints. Unpublished research paper from the USC Center for Robotics and Embedded Systems Lab., 2004

54. A. Ettlin, H. Bleuler, Randomised rough-terrain robot motion planning. in *Proceedings of the IEEE/RSJ International Conference on Intelligent Robots and Systems*, pp. 5798–5803 (2006)

55. D. Ferguson, A. Stentz, Anytime RRTs. in *Proceedings of the IEEE/RSJ International Conference on Intelligent Robots and Systems*, pp. 5369–5375 (2006)

56. S. Karaman, M.R. Walter, A. Perez, E. Frazzoli, S. Teller, Anytime motion planning using the RRT. in *Proceedings of the IEEE International Conference on Robotics and Automation*, pp. 1478–1483 (2011)

57. J.A. Primbs, V. Nevistic, J.C. Doyle, Nonlinear optimal control: a control Lyapunov function and receding horizon perspective. Asian J. Control **1**, 14–24 (1999)

58. T. Raff, C. Ebenbauer, F. Allgoewer, *Nonlinear Model Predictive Control: Passivity-based Approach* (Springer, New York, 2007)

59. A. Tahirovic, G. Magnani, P. Rocco, Mobile robot navigation using passivity-based MPC. in *Proceedings of the IEEE/ASME International Conference on Advanced Intelligent, Mechatronics*, pp. 248–488, July 2010

60. A. Tahirovic, G. Magnani, General framework for mobile robot navigation using passivity-based MPC. IEEE Trans. Autom. Control **56**(1), 184–190 (2011)

61. A. Tahirovic, G. Magnani, Passivity-based model predictive control for mobile robot navigation planning in rough terrains. in *Proceedings of the 2010 IEEE/RSJ International Conference on Intelligent Robots and Systems*, Oct 2010

62. C.A. Desoer, M. Vidyasagar, *Feedback Systems: Input-Output Properties* (Academic Press, New York, 1975)

63. J.C. Willems, Dissipative dynamical systems part 1.: general theory. Arch. Ration. Mech. Anal. **45**(5), 321–351 (1972)

64. J.C. Willems, Dissipative dynamical systems part 2.: linear systems with quadratic supply rates. Arch. Ration. Mech. Anal. **45**(5), 352–393 (1972)

65. D. Youla, L. Castriota, H. Carlin, Bounded real scattering matrices and the foundations of linear passive network theory. IRE Trans. Circ. Theory **6**(1), 102–124 (1959)

66. P. Moylan, B. Anderson, Nonlinear regulator theory and an inverse optimal control problem. IEEE Trans. Autom. Control **18**(5), 460–465 (1973)

67. D. Hill, P. Moylan, The stability of nonlinear dissipative systems. IEEE Trans. Autom. Control **21**(5), 708–711 (1976)

68. C.I. Byrnes, A. Isidori, J. Willems, Passivity, feedback equivalence, and the global stabilization of minimum phase nonlinear systems. IEEE Trans. Autom. Control **36**(11), 1228–1240 (1991)

69. A. Ailon, R. Ortega, An observer-based set-point controller for robot manipulators with flexible joints. Syst. Control Lett. **21**(4), 329–335 (1993)

70. H.K. Khalil, *Nonlinear Systems*, 3rd edn. (Prentice Hall, Upper Saddle River, 2002)

# Chapter 2
# PB/MPC Navigation Planner

## 2.1 Introduction

In this chapter, a rather straightforward procedure is presented to obtain navigation algorithms for a broad class of vehicle models, based on an adapted version of the passivity-based nonlinear MPC examined in [1]. The proposed PB/MPC approach for navigation planning can be seen as a generalization of the well-known DWA developed in [2–4]. Similar to the navigation based on the MPC/CLF [5], the PB/MPC optimization setup guarantees the task completion, which means the vehicle is being able to reach the goal position. However, whereas in the MPC/CLF navigation framework a control action that decreases the Lyapunov function has to be found in advance, which is rather difficult if not impossible for complex vehicle models, the PB/MPC navigation framework gives directly the control action as a consequence of the passivity-based control. Therefore, the PB/MPC can be easily adapted to a variety of vehicle and terrain models providing a straightforward procedure for the navigation of wide range of vehicles.

The first step of the procedure requires to shape the energy of the vehicle model by the navigation function which includes information on the goal position and obstacles. A navigation function is constructed for the field or terrain to be traveled in order to shape the energy of the vehicle model including the information on the goal position into the optimization setup. The second step is the selection of the output to force the system to be passive. Passivity-based control is used to make the system equilibrium point globally asymptotically stable, thus guaranteeing task completion.

The obtained properties of the PB/MPC navigation planner can be described as follows. The passivity-based constraint, inherently included in the PB/MPC optimization setup, enhances the navigation by guaranteeing the task completion. In accordance with the MPC paradigm, any additional constraint can be easily imposed into the optimization setup. A general vehicle model is extended by energy-shaping technique using the navigation function, which includes information on the terrain obstacles and the goal position, guaranteeing the avoidance of obstacles while

A. Tahirovic and G. Magnani, *Passivity-Based Model Predictive Control for Mobile Vehicle Motion Planning*, SpringerBriefs in Control, Automation and Robotics, DOI: 10.1007/978-1-4471-5049-7_2, © The Author(s) 2013

approaching the goal position. Unlike the MPC/CLF navigation framework, where a control action that decreases the value of Lyapunov function has to be found in advance, which is difficult if not impossible for complex vehicle models, the proposed PB/MPC framework gives a control action for any kind of vehicle and terrain models as a consequence of the passivity-based control. Therefore, the PB/MPC approach can be used also for mobile vehicles traveling in outdoor rough terrains, whose behavior is described by truly complex models.

## 2.2 PB/MPC Optimization Framework

The passivity-based nonlinear control approach, introduced in [1], has been exploited to propose a new mobile vehicle navigation framework for flat terrains [6, 7]. In addition, the PB/MPC navigation framework is extended from flat to rough terrains in [8]. The PB/MPC motion planning optimization framework can be described by the following optimization setup (2.1–2.9):

$$J(\mathbf{u}, \mathbf{r}(x_0)) = \int_{t_0}^{t_0+T} \gamma(\mathbf{x}, \mathbf{u}) dt + \Gamma(t_0 + T), \tag{2.1}$$

$$V(\mathbf{x}) = k\mathrm{NF}(\mathbf{r}) + \frac{1}{2} v^2, \tag{2.2}$$

$$\frac{d}{dt}\mathbf{x} = f(\mathbf{x}) + g(\mathbf{x})\mathbf{u} \tag{2.3}$$

$$\mathbf{y} = h(\mathbf{x}) = \left[\frac{\partial V}{\partial x} g(\mathbf{x})\right]^T, \tag{2.4}$$

$$\mathbf{u}^T(t)\mathbf{y}(t) < -\mathbf{y}^T(t)\phi(\mathbf{y}) \tag{2.5}$$

$$\tau : [0, 1] \rightarrow C_{\mathrm{free}}, \tau(0) = \mathbf{q}(t_0), \tau(1) = \mathbf{q}(t_0 + T) \tag{2.6}$$

$$v(t_0 + T) = 0 \tag{2.7}$$

$$\cos \angle(\nabla \mathrm{NF}, \mathbf{e}_{\hat{r}})|_{t=t_0+T_1} < 0 \\ \cos \angle(\nabla \mathrm{NF}, \mathbf{e}_{\hat{r}})|_{t=t_0+T} < 0 \tag{2.8}$$

$$\mathrm{NF}(\mathbf{r}(t_0 + T)) < \mathrm{NF}(\mathbf{r}(t_0 + T_1)) < \mathrm{NF}(\mathbf{r}(t_0)) \tag{2.9}$$

The main difference between the PB/MPC planners for flat and for rough terrains is in the cost function (2.1). In addition, the last constraint (2.9) is not necessarily required for the PB/MPC planner for flat terrains. These differences will be explained in the sequel to the chapter.

## *2.2.1 Cost Function*

The task of this optimization is to find such a control input $\mathbf{u}$ to guide the vehicle (traction force and steering angle momentum) for each optimization time horizon $t \in (t_0, t_0 + T)$, over all potential alternatives, by minimizing the cost function $J(\mathbf{u})$ given in (2.1). The integrand $\gamma(\mathbf{x}, \mathbf{u})$ is selected depending on what is locally required to minimize. The term $\Gamma(t_0 + T)$ represents an estimation of the cost-to-go value with respect to the goal position in terms of selected measure.

For the PB/MPC motion planner on flat terrains, the cost function is selected to be the value of the energy storage function at the end of the optimization horizon

$$J(\mathbf{u}) = V(\mathbf{x}(t_0 + T)), \tag{2.10}$$

where function $V(\mathbf{x})$ is selected as in (2.2). It includes a virtual potential term constructed by the navigation function of the given terrain, NF($\mathbf{r}$) [9, 10], which is selected to have a unique minimum at the goal position, and a kinetic term, $\frac{1}{2}v^2$, where $\mathbf{r} = (x_{cg}, y_{cg})$ and $v$ are the current vehicle coordinates and velocity. The concept of the navigation function has been rigorously introduced in [11]. The main motive behind the construction of such a function is the problem of the local minima which inherently appear in potential fields functions [12]. In [11], the authors have found the exact analytical expressions to construct a function which includes all terrain obstacles while having only a global minimum in the goal position. Some numerical solutions to the same problem has been given in [9, 10], where the navigation function might be expressed as being the shortest paths from each cell of a grid terrain to the goal position. The numerical navigation function given in [9] (NF1), is used in this work. The NF($\mathbf{r}$) function is computed for each point of a rectanguloid grid, which is made by appropriate terrain map discretization, as the $\mathcal{L}^1$ (Manhattan) distance to the goal position. In such a case, NF($\mathbf{r}$) function approximately represents the shortest path to the goal from each obstacle-free terrain point. This choice assures NF($\mathbf{r}$) has a unique minimum at the goal position. A numerical navigation function is not a differentiable function which might cause some problems while using algorithms to solve optimal control problems. The problem of differentiability of a numerical function will be addressed in Chap. 4 under Sect. 4.3.

Since the navigation function has a unique minimum at the goal position [9], the energy storage function given by (2.2) has a unique global minimum at $(x_{cg}\; y_{cg}\; v) = (x_{goal}\; y_{goal}\; 0)$. Hence, by decreasing the objective function (2.10), the vehicle gradually approaches the goal position by choosing the shortest possible paths while satisfying the PB/MPC optimization constraints (2.3–2.9). $V$ follows the definition of the CLF used in the MPC/CLF paradigm proposed in [5] for the particular problem of the navigation of the unicycle mobile robot in flat terrain.

However, choosing the shortest path to the goal position may be a rather strict constraint especially when the vehicle moves in rough terrains. A natural choice of the cost function (2.1) for the PB/MPC planner on rough terrains is a roughness measure computed along the path to be traversed. For this purpose, we introduce some possible cost functions that can describe the roughness level, or traversability index, along the selected path. The first candidate for a cost measure is a function that penalizes high roll and pitch values along the path. Such a function is used in [13] and is given by

$$\gamma(\mathbf{x}, \mathbf{u}) = (1 + \alpha(\varphi^2 + \theta^2)), \tag{2.11}$$

where $\varphi$, $\theta$ are the roll and pitch angles of the vehicle along the candidate path. Coefficient $\alpha$ represents the tradeoff between the minimum-time and minimum slope-dwell solutions. If $\alpha = 0$, then the solution gives the fastest path within the horizon.

A second candidate was proposed in [14] and is given by

$$\gamma(\mathbf{x}, \mathbf{u}) = \frac{1}{v_{\max}(\mathbf{r})}. \tag{2.12}$$

This function describes the roughness level in terms of the high mobility of the vehicle, where $v_{\max}(\mathbf{r})$ is the predicted maximal value of the vehicle velocity at each position $\mathbf{r}(t)$, $\forall t \in (t_0, t_0 + T)$, along a candidate path, which still does not cause sideslip and rollover of the vehicle [15, 16]. This function is more descriptive when it is important to increase the vehicle mobility. It favors those paths that allow high speeds while preserving the vehicle stability constraints.

Other possibilities for the estimation of the roughness measures along a candidate path are given in [17–19], where the authors introduced a traversability index describing the roughness of the terrain. Regardless of the choice of the measure, the vehicle prefers to find smoother regions toward the goal position, since all the aforementioned measures represent a kind of traversability index along a candidate path considered within the optimization.

For demonstration purposes, a local measure of the roughness is estimated by the relative height of the terrain describing its deviation from flatness and is given by

$$\gamma(\mathbf{x}, \mathbf{u}) = \frac{\sqrt{\mathrm{var}(z(\mathcal{R}))}}{d}, \tag{2.13}$$

where $d$ is the vehicle wheel diameter scaling the selected roughness measure to vehicle size, and $\sqrt{\mathrm{var}(z(\mathcal{R}))}$ being the standard deviation of the terrain height, $z(\mathcal{R})$, along a candidate path, where $\mathcal{R}$ is a terrain map [20]. This approximation is done for all candidate paths within the optimization horizon. However, the approach remains general since any aforementioned roughness functions can be used.

The proposed algorithm optimizes the roughness level toward the goal position in order to select smoother paths. To this purpose, the cost-to-go term $\Gamma(t_0 + T)$ representing the roughness-to-go value at the end of the optimization horizon, is added to the locally used roughness measure, in accordance with (2.1). The estimation of the optimal cost-to-go value within a nonlinear MPC framework is often impossible, and some rough estimations have to be found. Obtaining the optimal cost-to-go map for each terrain location is also likely impossible especially for the problems of the vehicle navigation on large-scale rough terrains. The reason is that the differential constraints have to be taken into account starting from each vehicle initial configuration. One way to construct a numerical roughness-based navigation function for the purpose of an MPC-like motion planning has been presented in [21, 22]. In addition, every time new information is acquired by the vehicle, the update procedure of the $\mathbf{D}^*$ algorithm [23, 24] can be used to get the updated cost-to-go map. Obtaining a differentiable objective function required to solve the optimization problem is addressed in Chap. 4 under Sect. 4.3.

Finally, the choice of the optimization and control horizon design parameters, $T$ and $T_1$, can influence the final result. The control horizon, $T_1$, can be a sample cycle period as in most MPC schemes, while the choice of $T$ can be further analyzed. However, regardless of the choice of these parameters, we assume only that $L_{max} \geq T \cdot v_{max}$, where $L_{max}$ is the maximum radius of the visible region with respect to the vehicle current position, and $v_{max}$ being the vehicle maximum velocity.

## 2.2.2 Optimization Constraints

Equation (2.3) represents the virtual model obtained by shaping the energy of the real vehicle dynamics by the navigation function, where $\mathbf{x}$ are the new states. The choice of the output (2.4) forces the system to be passive with respect to a radially unbounded and continuously differentiable storage function $V$, and is based on the passivity control concept.

Passivity constraint (2.5), where $\phi$ represents a damping injected to the model, asymptotically stabilizes the goal position providing the decrease of the energy storage function $V$.

Equation (2.6) constrains the optimization to the collision-free configurations, where $\tau$ is the map from the initial to the final vehicle configuration $\mathbf{q}$, into the collision-free space $C_{free}$. Constraint (2.7) guarantees that the selected control $\mathbf{u}$ can stop the vehicle at the end of the horizon satisfying collision-free constraint (2.6). If this constraint holds, then any state space point $\mathbf{x}(T_1)$ preserves the safe policy, where $T_1 < T$. The optimization preformed for the horizon $T$ repeats each $T_1$.

The conditions given in (2.8), where $\angle(\nabla \text{NF}, \mathbf{e}_{\dot{r}})$ is the current angle between the gradient of the navigation function and the current vehicle velocity direction $\dot{\mathbf{r}}$ represented by its unit vector $\mathbf{e}_{\dot{r}}$, are the terminal conditions which keep the vehicle oriented toward the decrease of the navigation function at the end of each PB/MPC optimization cycle. This constraint guarantees that the energy-shaped system includes some properties providing the asymptotic stability of the system [6].

The passivity constraint (2.5) ensures the decrease of the energy storage function $V$. This does not ensure the decrease of the virtual potential term NF (2.2) of $V$ at any time within the optimization horizon because of the kinetic energy term $\frac{1}{2}v^2$. As it was already discussed the choice of the objective function given in (2.10) yields the minimization of the value of the navigation function at the end of each optimization horizon. Such optimization policy aims at generating paths toward the steepest descent of the navigation function surface assuring the orientation of the vehicle to be toward its decrease. In case a cost function different from (2.10) is used, in such cases of rough terrains, an additional constraint must be imposed into the PB/MPC framework given by (2.9) to ensure the decrease of the navigation function along the selected paths. Since the optimization is performed within the time $T$ while the control action $\mathbf{u}$ is applied each time $T_1$, both conditions of (2.9) need to be included.

In addition to the constraints given by Eqs. (2.3–2.9), any additional constraints such as the control input limitations (e.g., maximum velocity), preventing the vehicle from the rolling over or from the slippage, can be easily included into the optimization setup.

### 2.2.3  Optimization Techniques

The MPC optimization can be conducted by a discrete number of motion primitives using the extreme-left, left, straight, right, and the extreme-right maneuvers which use the maximum acceleration allowed within a particular time horizon by the PB/MPC navigation scheme in accordance with the expression derived later (3.36). However, the MPC optimization problem can be solved by parameterization of the control space within the given horizon as it is nicely demonstrated in [13], when the optimization is solvable by a nonlinear programming optimization technique. Another approach based on a priori defined motion primitives that has widely been used in mobile vehicle navigation [25], was also used in [26]. In [6], the genetic algorithm (GA) is implemented for the local optimization as an alternative to the optimization approach based on defined motion primitives, where chromosomes consisted of the potential values of the vehicle accelerations and steering angles. Such optimizations provide more efficient ways of covering the vehicle control space improving the final solution. However, another way of tackling the local environment constraints that can be used instead of control space sampling is the vehicle state space sampling of feasible motions as discussed and illustrated in [27]. A reader can find a comprehensive overview of different optimization techniques used for the vehicle navigation in [27]. The possible implementation of an MPC-like motion planning algorithm using an optimal control software is presented in Chap. 4 under Sect. 4.3. Therein, a possible lack of a feasible solution for an optimization cycle is addressed by introducing an alternative motion strategy to keep the vehicle moving forward. Such a backup strategy can also be used as a candidate control by the GA algorithm in order to avoid the problem of not finding a feasible solution by the MPC-like optimization framework. Results suggest that both a GA and an optimal control software can be used to solve the optimization setup given by Eqs. (2.1–2.9).

As an example, Fig. 2.1 shows the block diagram of the PB/MPC motion planning approach for rough terrains. The 'Energy-shaped Virtual Model' and 'Constraint I: Passivity based' blocks are used to shape the energy of the system and to obtain a passive system, (2.2–2.5). The 'Constraint II: Steering' block provides the feasible set of steering angles (or steering momenta) $\mathbf{U}_s$, where $u_s \in \mathbf{U}_s$. The 'Constraint III:

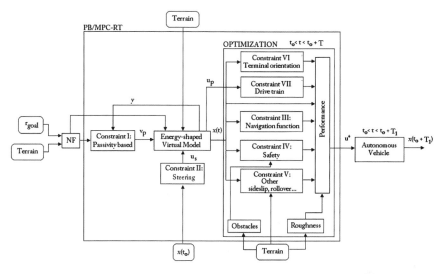

**Fig. 2.1** Block diagram of the PB/MPC navigation approach for rough terrains

**Navigation function'** and '**Constraint IV: Safety'** blocks check for the constraint on decrease of the navigation function (2.9) and the safety conditions (2.6, 2.7), respectively. The '**Constraint V'** block may include some additional constraints related to the vehicle stability conditions such as sideslip and rollover. These constraints estimate the maximum admissible vehicle velocity profile along a candidate path that still does not cause sideslip or rollover of the vehicle. The passivity-based constraint (2.5) provides the feasible set $\mathbf{V}_p$ for the new traction force input $v_p \in \mathbf{V}_p$. The traction force limitations of the input $u_p$ can be checked using the block '**Constraint VII: Drive Train'** within the optimization. Finally, the block: '**Constraint VI: Terminal Orientation'** is used to force the vehicle to be oriented toward the decrease of the navigation function at the end of each MPC optimization horizon in accordance with (2.8).

## 2.3 Design of the PB/MPC Motion Planner

### 2.3.1 General Model

The model of a mobile vehicle driven with the traction force $u_p$ and the steering angle momentum $u_s$, can be expressed in a rather general form by [28]

$$\dot{x} = f(x) + G(x)(u_p \ u_s)^T, \tag{2.14}$$

where $x \in R^n$ is the vehicle state vector, $x = (x_{cg} \; y_{cg} \; v \; x_4 \; x_5 \; \ldots \; x_n)^T$ and the position of the vehicle center of mass $(x_{cg}, y_{cg})$, and the velocity of the vehicle $v$ are the states of interest with respect to the final goal position. $(x_4 \; x_5 \; \ldots \; x_n)$ are the remaining states (see Chap. 3). The goal of the navigation task can be stated as making $(x_{cg} \; y_{cg} \; v)_e = (x^* \; y^* \; 0)$ to be a globally asymptotically stable equilibrium point, being $(x^*, y^*)$ the goal position and $(\cdot)_e$ denoting the equilibrium point of the corresponding subsystem.

In order to have the goal position included in a candidate state for the zero equilibrium point of the subsystem represented by $(x_{cg} \; y_{cg} \; v)$, the position coordinates have to be transformed as $e_x = x_{cg} - x^*$, $e_y = y_{cg} - y^*$, yielding

$$\dot{e} = f(e) + G(e)(u_p \; u_s)^T, \tag{2.15}$$

where

$$e = (e_x \; e_y \; v \; x_4 \; x_5 \ldots x_n)^T,$$

$$f(e) = (v f_{e_x} \; v f_{e_y} \; f_v \; f_{x_4} \; f_{x_5} \; \cdots \; f_{x_n})^T,$$

$$G(e) = \begin{pmatrix} g_{e_x 1} \; g_{e_y 1} \; g_{v1} \; g_{41} \; g_{51} \; \cdots \; g_{n1} \\ g_{e_x 2} \; g_{e_y 2} \; g_{v2} \; g_{42} \; g_{52} \; \cdots \; g_{n2} \end{pmatrix}^T.$$

It can be seen that the first two equations in (2.15) carry the information on the vehicle nonholonomic constraint.

## 2.3.2 Energy-Shaping Using a Navigation Function

In general, the equilibrium point of the subsystem described by $(e_x \; e_y \; v)$ will not be the zero one, that is $(e_x \; e_y \; v)_e \neq (0 \; 0 \; 0)$, since the system has no global information about the goal position. The traction force input $u_p$ can be selected to shape the energy of the system with an injection of additional information on the exact goal position providing $(e_x \; e_y \; v)_e = (0 \; 0 \; 0)$ for this purpose. This can be achieved using a navigation function of the given terrain $NF(\mathbf{r})$ with a unique minimum at the goal position, that is $\min NF(\mathbf{r}) = NF(\mathbf{r}^*)$, as the part of the selected traction force input. Many different ways to compute NF are discussed in [9].

It is worth noting that

$$\|\dot{\mathbf{r}}\| = \sqrt{\dot{x}_{cg}^2 + \dot{y}_{cg}^2} = \sqrt{\dot{e}_x^2 + \dot{e}_y^2} = v f_r, \tag{2.16}$$

where $f_r = \sqrt{f_{e_x}^2 + f_{e_y}^2}$ is a nonnegative function, since for a general vehicle model and for some $\mathbf{r}$ it may result $f_r \neq 1$, implying $\|\dot{\mathbf{r}}\| \neq v$.

It is assumed that $g_{v1}$ is a nonsingular and positive function for all $x \in R^n$, and that the states of system (2.15) are measured or estimated at the end of each optimization horizon, since the MPC optimization requires the initial values of the states of the reshaped vehicle model for the subsequent optimization cycle. Besides vehicle velocity $v$, and steering angle $\delta$, usually a SLAM has to be solved to obtain the vehicle position $\mathbf{r}$, and its heading angle $\psi$. Other vehicle states can be measured by the Inertial Measurement Units and other sensors. In this case, the energy of the system can be shaped taking

$$u_p = g_{v_1}^{-1}(-f_v - k\nabla \text{NF}(\mathbf{r})\mathbf{e}_{\hat{r}} f_r - g_{v_2} u_s) + v_p, \tag{2.17}$$

where

$$k\nabla \text{NF}(\mathbf{r})\mathbf{e}_{\hat{r}} = k\|\nabla \text{NF}(\mathbf{r})\| \cos \angle(\nabla \text{NF}(\mathbf{r}), \mathbf{e}_{\hat{r}}) \tag{2.18}$$

is the scaled inner product of the gradient of the navigation function $\text{NF}(\mathbf{r})$ with a unit vector of the current direction represented by vector $\mathbf{e}_{\hat{r}}$. This term favors those directions that move the vehicle along the paths which decrease the value of the navigation function $\text{NF}(\mathbf{r})$, hence toward the goal position. For instance, if the vehicle goes in the direction of the steepest descent of the navigation function, (2.18) will have a minimum possible value thus providing the maximum value in (2.17). If the vehicle moves increasing values of the navigation function, the component (2.18) will have positive values, hence decreasing the speed and stopping the vehicle. $v_p$ is the new control input of the system replacing the traction force.

In [4] the authors discussed how to construct the navigation function for the case when the vehicle only has information about its sensor range environment.

Given (2.17), the system (2.15) transforms as

$$\dot{e} = \tilde{f}(e) + \tilde{G}(e)(v_p \; u_s)^T, \tag{2.19}$$

where

$$\tilde{f}(e) = (v\tilde{f}_{e_x} \; v\tilde{f}_{e_y} \; \tilde{f}_v \; \tilde{f}_{x_4} \; \tilde{f}_{x_5} \; \cdots \; \tilde{f}_{x_n})^T,$$

$$\tilde{G}(e) = \begin{pmatrix} g_{e_x 1} \; g_{e_y 1} \; g_{v1} \; g_{41} \; \cdots \; g_{n1} \\ \tilde{g}_{e_x 2} \; \tilde{g}_{e_y 2} \; \tilde{g}_{v2} \; \tilde{g}_{42} \; \cdots \; \tilde{g}_{n2} \end{pmatrix}^T,$$

and defining

$$g_{11} = g_{e_x 1}; \; g_{21} = g_{e_y 1}; \; g_{31} = g_{v1}; \quad f_{x1} = vf_{e_x}; \; f_{x2} = vf_{e_y};$$
$$g_{12} = g_{e_x 2}; \; g_{22} = g_{e_y 2}; \; g_{32} = g_{v2}, \quad f_{x3} = f_v,$$

it results

$$\tilde{g}_{i2} = g_{i2} - g_{i1}\frac{g_{v2}}{g_{v1}}, \tag{2.20}$$

$$\tilde{f}_{xi} = f_{xi} + \frac{g_{i1}}{g_{v1}} \left( -f_v - k\nabla NF(\mathbf{r})\mathbf{e}_{\hat{r}} f_r \right), i = 1, \dots, n; \tag{2.21}$$

providing $\tilde{g}_{v_2} \equiv 0$, and

$$\dot{v} = -k\nabla NF(\mathbf{r})\mathbf{e}_{\hat{r}} f_r + g_{v_1} v_p. \tag{2.22}$$

For the sake of consistency of the notation to the passivity based control theory presented in the next section, (2.19) will be denoted as

$$\dot{x} = f(x) + G(x) \begin{pmatrix} v_p \\ u_s \end{pmatrix}. \tag{2.23}$$

**Theorem 2** *If the general vehicle dynamic model (2.14) is transformed into (2.15) and the system energy is shaped by (2.17), the subsystem described by the states $x_{ss} = (e_x \ e_y \ v)^T$ of the new vehicle dynamic model (2.19) will have zero equilibrium point, that is $x_{ss} = 0$.*

*Proof* Assuming $\dot{e}_x = 0$, $\dot{e}_y = 0$ and $\dot{v} = 0$, namely that there is no movement in both directions of the reference frame, the vehicle velocity is equal to zero, $v = 0$. From $\dot{v} = 0$, using (2.22) with input $v_p \equiv 0$, it follows $\nabla NF(\mathbf{r})\mathbf{e}_{\hat{r}} = 0$. One possible solution of the latter equation, $\cos(\nabla NF(\mathbf{r})\mathbf{e}_{\hat{r}})) = 0$, implies that the first condition in (2.8) is not satisfied for the given assumptions since this equality then also holds at the end of the operating time horizon $T_1$. This means that the feasible solution to this equation is only $\mathbf{r} = (x^* \ y^*)$, that is, $e_x = 0$ and $e_y = 0$, since the unique minimum of the navigation function $NF(\mathbf{r})$ is at this point.

### 2.3.3 Energy Storage Function

The main task of the navigation planning algorithm is to generate reference trajectories using the energy shaped model of the vehicle, whose subsystem consisting of the states of interest has an equilibrium point $x_{ss} = 0$. In order to find the inputs that move the vehicle toward the goal position, an appropriate energy storage function has to be selected. In [5] the authors used the direct Control Lyapunov function approach in order to navigate the unicycle mobile vehicle. The MPC/CLF optimization setup has the explicit constraint on the decrease of Lyapunov function along trajectories of the system, that is, $\dot{V} \leq -\epsilon\sigma(x)$, where the function $\epsilon\sigma(x)$ is a properly selected positive definite function. In order to satisfy this constraint, one should find such control action to make the system stable. For a unicycle mobile vehicle, the stabilizing control actions can be easily found as illustrated in [5]. Although there are some procedures to obtain stabilizing control actions using given Lyapunov function for particular class of systems, this is generally a hard task for many nonlinear systems.

Since in (2.19) a potential energy term exists in the form of the navigation function $NF(\mathbf{r})$, a natural choice of the energy storage function is similar to the one proposed in [5] and given by (2.2). A major difference with respect to the Lyapunov function proposed in [5], where only the unicycle mobile robot was considered, is that in (2.2) a distinction is made between $\|\dot{\mathbf{r}}\|$ and $v$, which is necessary for more accurate car-like vehicle models (see Eq. (2.16)). With the assumption that the navigation function $NF(\mathbf{r})$ can be constructed such that it is continuously differentiable, and since this storage energy function $V(x)$ is clearly radially unbounded and positive definite, this choice of $V(x)$ satisfies condition (1.5).

From (2.16) and (2.22), it results

$$\frac{\partial V}{\partial x} f(x) = \left( \frac{\partial V}{\partial \mathbf{r}} \frac{\partial V}{\partial v} \frac{\partial V}{\partial x_3} \cdots \frac{\partial V}{\partial x_n} \right) f(x) = (k \nabla NF(\mathbf{r}) \ v \ \mathbf{0}) f(x) \quad (2.24)$$

$$= k \nabla NF(\mathbf{r}) \dot{\mathbf{r}} + v(-k \nabla NF(\mathbf{r}) \mathbf{e}_{\dot{\mathbf{r}}} f_r) = 0, \forall x.$$

### 2.3.4 Passivity

In order to have a passive system, the output of the system is selected according to (2.4) and is given by

$$\mathbf{y}^T = \left( \frac{\partial V}{\partial \mathbf{r}} \frac{\partial V}{\partial v} \frac{\partial V}{\partial x_3} \cdots \frac{\partial V}{\partial x_n} \right) G(x) = (k \nabla NF(\mathbf{r}) \ v \ \mathbf{0}) G(x) \Rightarrow$$

$$\mathbf{y}^T = (g_{v_1} v \ 0) \Rightarrow y_1 = g_{v_1} v. \quad (2.25)$$

This means that the output of interest for the given storage function is $y_1$ that will be used to define a damping feedback to the model, as discussed in subsection 2.3.6, only through input $u_p$.

### 2.3.5 Zero State Observability

**Theorem 3** *If a general vehicle dynamic model (2.14) is transformed into (2.15) and if its energy shaping form and output are selected as in (2.17) and (2.3.4), respectively, the subsystem described by the three states of interest $x_{ss}$ of the new vehicle dynamic model (2.19) will be zero-state observable.*

*Proof* This claim is easy to verify starting from the subsystem

$$\begin{pmatrix} \dot{e}_x \\ \dot{e}_y \\ \dot{v} \end{pmatrix} = \begin{pmatrix} v \tilde{f}_{e_x} \\ v \tilde{f}_{e_y} \\ \tilde{f}_v \end{pmatrix} + \begin{pmatrix} g_{e_x 1} \ \tilde{g}_{e_x 2} \\ g_{e_y 1} \ \tilde{g}_{e_y 2} \\ g_{v1} \ \tilde{g}_{v2} \end{pmatrix} \begin{pmatrix} v_p \\ u_s \end{pmatrix}, \quad (2.26)$$

with the condition $y \equiv 0$ and $u \equiv 0$. This implies $v \equiv 0$ since the output is selected as in (2.3.4). This means $\dot{e}_x = 0$ and $\dot{e}_y = 0$, as well as $\dot{v} = 0$. Using (2.22), it follows $\nabla \mathrm{NF}(\mathbf{r})\mathbf{e}_{\dot{r}} = 0$. Similar to the discussion given in the proof of Theorem 2, the latter equation implies $\mathbf{r} = (x^* \ y^*)$, that is $e_x = 0$ and $e_y = 0$, so that $x_{ss} = 0$, namely the subsystem (2.26) is ZSO.

## 2.3.6 Stability

Since all conditions of Theorem 1 are satisfied, the input $v_p$ can be selected in the form $v_p = -\phi(y)$, where $\phi$ is any locally Lipschitz function such that $\phi(0) = 0$ and $y^T \phi(y) > 0$ for all $y \neq 0$. One possible choice of a damping injection using the function $\phi(y)$ is (see e.g. [29])

$$v_p = -\epsilon \frac{1}{g_{v_1}} \frac{2}{\pi} \arctan(k_v v), \tag{2.27}$$

where $\epsilon$ and $k_v$ are positive constants to be selected.

In order to obtain stability, by the assumption $v \geq 0$, one can write

$$v_p \leq -\epsilon \frac{1}{g_{v_1}} \frac{2}{\pi} \arctan(k_v v). \tag{2.28}$$

This choice of $v_p$ satisfies (2.5) making the equilibrium point $x_{ss}$ of subsystem (2.26), globally asymptotically stable.

This claim can be easily verified finding the time derivative of the energy storage function along the trajectories of system (2.19) under condition (2.28). In fact, using (2.22)

$$\dot{V} = \left( \frac{\partial V}{\partial \mathbf{r}} \ \frac{\partial V}{\partial v} \ \frac{\partial V}{\partial x_3} \ \cdots \ \frac{\partial V}{\partial x_n} \right) \dot{x} = (k \nabla \mathrm{NF}(\mathbf{r}) \ v \ \mathbf{0}) \dot{x}$$

$$= k \nabla \mathrm{NF}(\mathbf{r})\dot{\mathbf{r}} + v(-k \nabla \mathrm{NF}(\mathbf{r})\mathbf{e}_{\dot{r}} f_r + g_{v_1} v_p) = g_{v_1} v v_p$$

and, under constraint (2.28), the condition on the derivative of the energy storage function along the trajectories of the closed-loop system is obtained as follows:

$$\dot{V} = g_{v_1} v v_p \leq -\epsilon v \frac{2}{\pi} \arctan(k_v v). \tag{2.29}$$

Hence, $\dot{V}$ is negative semidefinite and $\dot{V} = 0$ if and only if $v = 0$. Then, by zero-state observability, $y(t) \equiv 0 \Rightarrow x_{ss} \equiv 0$. Therefore, according to the invariance principle, the origin of the subsystem represented by $x_{ss}$ is globally asymptotically stable. This means that the energy of the system under control, expressed by the sum of the virtual potential NF($\mathbf{r}$) and of the kinetic energy term $\frac{1}{2}v^2$, will decrease along the trajectories of the controlled system, moving the vehicle toward the goal position.

This result gives the final shape of the traction force $u_p$ given by (2.17) and (2.28), over each MPC optimization horizon.

# References

1. T. Raff, C. Ebenbauer, F. Allgoewer, *Nonlinear Model Predictive Control: Passivity-based Approach* (Springer, Berlin, 2007)
2. R. Simmons, The curvature-velocity method for local obstacle avoidance, in *Proceedings of the IEEE International Conference on Robotics and Automation*, 1996, pp. 3375–3382
3. D. Fox, W. Burgard, S. Thrun, The dynamic window approach to collision avoidance. IEEE Robot. Autom. Mag. **4**, 23–33 (1997)
4. O. Brock, O. Khatib, High-speed navigation using the global dynamic window approach, in *Proceedings of the IEEE International Conference on Robotics and Automation*, vol. 1. 1999, pp. 341–346
5. P. Oegren, N.E. Leonard, A convergent dynamic window approach to obstacle avoidance. IEEE Trans. Robot. **21**(2), 188–195 (2005)
6. A. Tahirovic, G. Magnani, P. Rocco, Mobile robot navigation using passivity-based MPC, in *Proceedings of the IEEE/ASME International Conference on Advanced Intelligent, Mechatronics*, July 2010, pp. 248–488
7. A. Tahirovic, G. Magnani, General framework for mobile robot navigation using passivity-based MPC. IEEE Trans. Autom. Control **56**(1), (2011)
8. A. Tahirovic, G. Magnani, Passivity-based model predictive control for mobile robot navigation planning in rough terrains, in *Proceedings of the 2010 IEEE/RSJ International Conference on Intelligent Robots and Systems*, Oct 2010
9. J. Latombe, *Robot Motion Planning* (Kluwer, Boston, 1991)
10. J. Barraquand, B. Langlois, J. Latombe, Numerical potential field techniques for robot path planning. IEEE Trans. Syst. Man Cybern. **22**(2), 224–241 (1992)
11. E. Rimon, D.E. Koditschek, Exact robot navigation using artificial potential fields. IEEE Trans. Robot. Autom. **8**, 501–518 (Oct. 1992)
12. O. Khatib, Real-time obstacle avoidance for manipulators and mobile robots. Int. J. Robot. Res. **5**(1), 90–98 (1986)
13. T.M. Howard, A. Kelly, Optimal rough terrain trajectory generation for wheeled mobile robots. Int. J. Robot. Res. **26**(2), 141–166 (2007)
14. K. Iagnemma, S. Shimoda, Z. Shiller, Near-optimal navigation of high speed mobile robots, in *Proceedings of the IEEE/RSJ International Conference on Intelligent Robots and Systems*, vol. 2. 2008, pp. 22—26
15. Z. Shiller, Y.-R. Gwo, Dynamic motion planning of autonomous vehicles. IEEE Trans. Robot. Autom. **7**(2), 241–249 (1991)
16. Z. Shiller, Obstacle traversal for space exploration, in *Proceedings of the IEEE International Conference on Robotics and Automation*, 2000
17. S. Singh, R. Simmons, T. Smith, A. Stentz, V. Verma, A. Yahja, K. Schwehr, Recent progress in local and global traversability for planetary rovers, in *Proceedings of the IEEE International Conference on Robotics and Automation*, 2000
18. H. Seraji, Fuzzy traversability index: a new concept for terrain-based navigation. J. Robot. Syst. **17**(2), 75–91 (2000)
19. A. Howard, H. Seraji, Vision-based terrain characterization and traversability assessment. J. Robot. Syst. **18**(10), 577–587 (2001)
20. K. Iagnemma, S. Dubowsky, *Mobile Robots in Rough Terrain: Estimation, Motion Planning and Control with Application to Planetary Rovers* (Springer, Berlin, 2004)

21. A. Tahirovic, G. Magnani, A roughness-based RRT for mobile robot navigation planning, in *Proceedings of the 18th IFAC World Congress*, 2011, pp. 5944–5949
22. A. Tahirovic, G. Magnani, Y. Kuwata, An approximate of the cost-to-go map on rough terrains, in *Proceedings of the IEEE International Conference on Mechatronics*, 2013
23. A. Stentz, The focussed D* algorithm for real-time replanning, in *Proceedings of the International Joint Conference on Artificial Intelligence*, pp. 1652–1659, 1995
24. S. Koenig, M. Likhachev, Fast replanning for navigation in unknown terrain. IEEE Trans. Robot. **21**(3), 354–363 (2005)
25. A. Lacaze, Y. Moscovitz, N. Declaris, K. Murphy, Path planning for autonomous vehicle driving over rough terrain, in *Proceedings of the IEEE International Symposium on Intelligent Control* (Gaithersburg, MD, 1998), pp. 50–55
26. M. Pivtoraiko, R.A. Knepper, A. Kelly, Differentially constrained mobile robot motion planning in state lattices. J. Field Robot. **26**(3), 308–333 (2009)
27. T.M. Howard, C.J. Green, A. Kelly, D. Ferguson, State space sampling of feasible motions for high-performance mobile robot navigation in complex environments. J. Field Robot. **25**(10), 325–345 (2008)
28. Y. Yoon, J. Shin, H.J. Kim, Y. Park, S. Sastry, Model-predictive active steering and obstacle avoidance for autonomous ground vehicles. Control Eng. Pract. **17**(7), 741–750 (2009)
29. H.K. Khalil, *Nonlinear Systems*, 3rd edn. (Prentice Hall, Upper Saddle River, 2002)

# Chapter 3
# Examples

## 3.1 Introduction

This chapter demonstrates the design procedure of the PB/MPC motion planning framework. The first two examples consider the vehicle models that might be used on flat terrains, a unicycle, and a car-like mobile vehicle. The third example covers a rather general model that can be used for rough terrains.

## 3.2 Flat Terrain

### 3.2.1 Unicycle Vehicle

The model of a unicycle vehicle with nonholonomic constraints can be written in the general form (2.14) with

$$\dot{x} = (\dot{x}_{cg} \ \dot{y}_{cg} \ \dot{v} \ \dot{\psi} \ \ddot{\psi})^T,$$
$$f(x) = (vf_{e_x} \ vf_{e_y} \ 0 \ \dot{\psi} \ 0)^T,$$

$$G(x) = \begin{pmatrix} 0 & 0 & \frac{1}{m} & 0 & 0 \\ 0 & 0 & 0 & 0 & \frac{1}{I_{zz}} \end{pmatrix}^T,$$

where $\psi$, $m$, and $I_{zz}$ are the current vehicle orientation with respect to the given reference frame, the vehicle mass, and the yaw moment of inertia, respectively (see e.g., [1]).

After the coordinate transformation $e_x = x_{cg} - x^*$, $e_y = y_{cg} - y^*$, the form (2.15) is obtained with $e = (\dot{e}_x \ \dot{e}_y \ \dot{v} \ \dot{\psi} \ \ddot{\psi})^T$, $f(e) = f(x)$, $f_{e_x} = \cos\psi$, $f_{e_y} = \sin\psi$ and $G(e) = G(x)$. Here, relation (2.16) is $\|\dot{\mathbf{r}}\| = v$, hence $f_r \equiv 1$.

The energy of the system is shaped with the traction force input given in (2.17)

A. Tahirovic and G. Magnani, *Passivity-Based Model Predictive Control for Mobile Vehicle Motion Planning*, SpringerBriefs in Control, Automation and Robotics, DOI: 10.1007/978-1-4471-5049-7_3, © The Author(s) 2013

$$u_p = m(-k\nabla NF(\mathbf{r})\mathbf{e}_{\hat{r}}) + v_p. \tag{3.1}$$

Considering expressions (2.20) and (2.21), the system

$$\tilde{f}(e) = (v\cos\psi \quad v\sin\psi \quad -k\nabla NF(\mathbf{r})\mathbf{e}_{\hat{r}} \quad \dot{\psi} \quad 0)^T,$$
$$\tilde{G}(e) = G(e)$$

is obtained.

As it has already been shown, this system satisfies all conditions of Theorem 1 with respect to the storage energy function (2.2). This means that with the output choice $y = \frac{1}{m}v$ (2.25), for any traction force input selected from the set given in (2.28) $v_p \le -\epsilon m \frac{2}{\pi}\arctan(k_v v)$, (2.29) holds and the state point $x_{ss} = 0$ is globally asymptotically stable.

This result gives the final shape of the traction force input within each MPC optimization horizon similar to the one obtained in [1], where it has been computed using the MPC/CLF approach.

### 3.2.2 Car-Like Mobile Vehicle with Slippage

Using the tyre model described by Pacejka's MAGIC formula [2, 3], the model of the car-like vehicle with slippage can be described according to the general form (2.14) with (see e.g., [4])

$$\dot{x} = (\dot{x}_{cg} \ \dot{y}_{cg} \ \dot{v} \ \dot{\psi} \ \ddot{\psi} \ \dot{\beta} \ \dot{\delta})^T,$$

$$f(x) = (vf_{e_x} \ vf_{e_y} \ f_v \ \dot{\psi} \ f_{\dot{\psi}} \ f_\beta \ -\frac{1}{\tau_s}\delta)^T,$$

$$G(x) = \begin{pmatrix} 0 & 0 & g_{v_1} & 0 & 0 & g_{\beta_1} & 0 \\ 0 & 0 & 0 & 0 & 0 & 0 & \kappa_s \end{pmatrix}^T, \tag{3.2}$$

where $\psi$, $\delta$, and $\beta$ are the current vehicle orientation with respect to the given reference frame, the steering angle, and the velocity direction with respect to the vehicle reference frame, respectively (Fig. 3.1).

After coordinate transformation the form (2.15) is obtained with

$$e = (\dot{e}_x \ \dot{e}_y \ v \ \psi \ \dot{\psi} \ \beta \ \delta)^T,$$
$$f(e) = f(x), \tag{3.3}$$
$$G(e) = G(x), \tag{3.4}$$

where

**Fig. 3.1** A schematic diagram
of the vehicle model with slip
angle

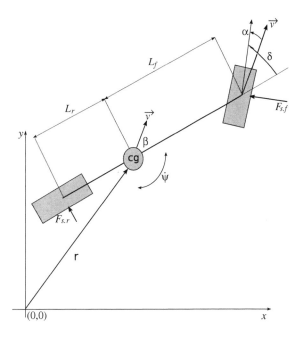

$$f_{e_x} = \cos\beta\cos\psi - \sin\beta\sin\psi, \; f_{e_y} = \cos\beta\sin\psi + \sin\beta\cos\psi. \tag{3.5}$$

$f_v$ is obtained from the longitudinal motion dynamics equation

$$m\dot{v} = F_x\cos\beta + F_y\sin\beta, \tag{3.6}$$

$F_x$ and $F_y$ being the forces acting in the $x$ and $y$ directions of the vehicle internal reference frame, respectively, given by

$$F_x = [-2\sin\delta F_{s.f} + u_p], \; F_y = [2\cos\delta F_{s.f} + 2F_{s.r}]. \tag{3.7}$$

$F_{s.f}$ and $F_{s.r}$ are the lateral forces acting on the front and rear wheels, respectively. These forces can be approximated using the stiffness coefficients $C_f$ and $C_r$, as $F_f = C_f\alpha_f$ and $F_r = C_r\alpha_r$, $\alpha_f$ and $\alpha_r$ being the slip angles of the front and rear tyres approximated by

$$\alpha_f = \delta - \arctan\frac{v\sin\beta + L_f\dot{\psi}}{v\cos\beta}, \; \alpha_r = -\arctan\frac{v\sin\beta - L_r\dot{\psi}}{v\cos\beta}, \tag{3.8}$$

where $L_f$ and $L_r$ are the distances of the front and rear wheels from the vehicle center of mass, respectively.

$f_\beta$ is obtained from the momentum equation

$$mv(\dot{\beta} + \dot{\psi}) = -F_x \sin \beta + F_y \cos \beta. \tag{3.9}$$

Since $F_x$ is included in (3.9), which depends on the traction force input $u_p$, it turns out that the coefficient $g_{\beta_1} \neq 0$ in $G(x)$.

$f_{\dot{\psi}}$ is obtained from the momentum equation

$$I_{zz}\ddot{\psi} = 2F_{s,f}L_f \cos \delta - 2F_{s,r}L_r. \tag{3.10}$$

Here, relation (2.16) is $\|\dot{\mathbf{r}}\| = v$, hence $f_r \equiv 1$. From (3.6) and (3.7) it follows that $g_{v_1} = \frac{1}{m} \cos \beta$, and $\beta$ being small, $\cos \beta \approx 1$, the function $g_{v_1}$ is nonsingular. Therefore, the energy of the system can be shaped with the traction force input given in (2.17)

$$u_p = \frac{m}{\cos \beta}(-f_v - k\nabla \mathrm{NF}(\mathbf{r})\mathbf{e}_{\dot{r}}) + v_p, \tag{3.11}$$

where $f_v$ is extracted from (3.6).

As it has been already shown, this energy reshaped system fulfills all conditions of Theorem 1 with respect to the storage energy function (2.2). This means that with the output choice $y = \frac{\cos \beta}{m}v$ (2.25) and the traction force input satisfying (2.28) $v_p \leq -\epsilon\frac{m}{\cos \beta}\frac{2}{\pi} \arctan(k_v v)$, (2.29) holds and the state $x_{ss} = 0$ is globally asymptotically stable.

### 3.2.3 Simulations

The PB/MPC navigation framework was verified using the unicycle mobile vehicle, which has been also exploited in [1, 5], and the car-like vehicle with $v_{max} = 1.2$ m/s.

For the unicycle vehicle, the MPC optimization within one time horizon was carried out using a discrete number of vehicle maneuvers, that is the extreme-left, left, straight, right, and the extreme-right. This approach is demonstrated in Fig. 3.2a where all possible eligible paths have been evaluated at the beginning of each MPC time horizon while searching for the best one. All these maneuvers used the maximum acceleration allowed within a particular time horizon by the PB/MPC navigation scheme in accordance with (2.28). The main drawback of this limited set of motion primitives along the optimization horizon regard the existence of the feasible solution. If this simplified approach is used then it is important to select the time horizons $T_1$ and $T$ appropriately. In the simulation presented, values $T_1 = 0.3$s and $T = 0.8$s have provided sufficient reliability for different terrain examples.

Figure 3.2b illustrates a terrain example, where the vehicle was able to reach the goal position. Grazing the obstacles is a direct consequence of the selected construction algorithm of the navigation function (so-called NF1) (see [6]). It can be easily avoided by enlarging obstacles or using other navigation functions [1, 6]. The NF1 function is computed for each point of a rectangloid grid, which is made by

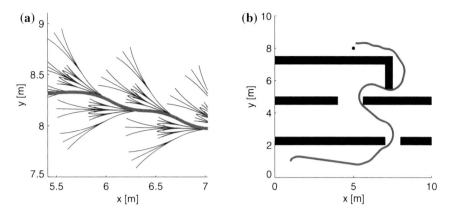

**Fig. 3.2** Unicycle vehicle: **a** Illustration of the simple optimization approach with a discrete number of maneuvers. **b** Generated path within the given configuration

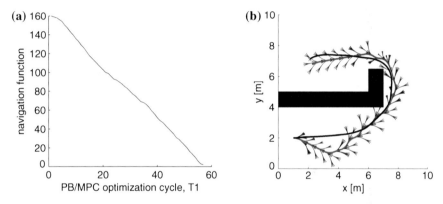

**Fig. 3.3** **a** Values of the navigation function at the end points of PB/MPC optimization cycles. **b** This figure shows that more efficient path can be obtained if more careful local optimization is used. Path presented with *black color* and no lattices was obtained by GA algorithm while the other one by the optimization with a few motion primitives

appropriate terrain map discretization, as the $\mathcal{L}^1$ (Manhattan) distance to the goal position. In such a case, NF1 function represents approximately the shortest path to the goal from each obstacle-free terrain point. This choice assures NF($\mathbf{r}$) has a unique minimum at the goal position. Figure 3.3a verifies the decrease of the navigation function from the initial position to the goal position guaranteeing the convergence of the algorithm. In the simulation, $k$ is set to 0.7 in order to obtain the final form of the virtual potential term, $kNF$.

Different optimizations can be used in order to find more efficient paths. Figure 3.3b illustrates a possible improvement based on a different local optimization technique. In this figure, the optimization based on the genetic algorithm (GA) is shown as a possible alternative, where a chromosome consists of control inputs,

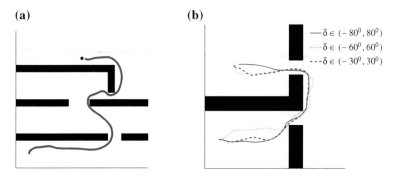

**Fig. 3.4** Car-like vehicle: **a** Generated path within the given terrain. **b** Paths generated using models with different steering constraints $\delta$

vehicle steering angle, and acceleration. It can be observed that GA can improve the result by covering the vehicle control space efficiently and giving a smoother path toward the goal position. GA was also used for the car-like vehicle, and the results obtained from models with different steering constraints are shown in Fig. 3.4.

The dumping injection is given by (2.28). If a large enough $k_v$ is taken, then $\frac{2}{\pi}\arctan(k_v v)$ can be considered a unit Heaviside function for $v \geq 0$. This choice is to make the PB/MPC optimization framework behave as the MPC/CLF for the navigation of a unicycle vehicle. Therefore, the PB/MPC optimization framework can be seen as a generalization of the MPC/CLF for this vehicle, also covering a wider class of vehicles and terrain models. The various dumping injections presented in [7] can also be used. Parameter $\varepsilon$ in (2.28) is related to the speed of decrease of the energy storage function $V$ (2.29) and is arbitrarily set to $\varepsilon = 0.1$.

## 3.3 Rough Terrains

The PB/MPC planner for rough terrains easily accommodates a thorough vehicle model that accounts for complex vehicle dynamics, terrain structure, and wheel-terrain interaction. In this way, the generated trajectories are likely to be feasible for rough terrains, unlike the trajectories obtained using a flat terrain vehicle model. In order to get feasible trajectories, the latter approach uses a feedback loop to compensate for the effects caused by terrain irregularities. However, the flat terrain model approach might lack feasible trajectories during the optimization.

The presented PB/MPC navigation approach has a compact optimization form which consists of global- and local-based planning strategies. Consequently, the approach guarantees the safe task completion. The PB/MPC is analyzed for various worst-case scenarios, providing insights into its strength and limitations in terms of the sensitivity to terrain roughness. The analysis includes the equations to compute the final completion time, the path length and its shape obtained during the task execution. The worst case possible shapes are derived explicitly for three possible cases, unknown rough terrain with obstacles, completely known rough terrain with obstacles, and unknown rough terrain without obstacles.

The main task of this section is to derive a virtual model of the vehicle that will be used to generate feasible trajectories by shaping the energy of the vehicle model in rough terrains. The energy-shaping technique is implemented by the navigation function constructed for the given terrain, while the goal position is asymptotically stabilized by the passivity-based control approach.

### 3.3.1 General Model of a Vehicle in Rough Terrain

The challenge in deriving the model of a vehicle acting in a rough terrain is to include the vehicle longitudinal and lateral dynamics, the suspension and tyre compliance and the terrain properties and uncertainties (see, e.g., [2, 3, 8]). The aim of this subsection is to comprise the most important effects into the nonlinear vehicle model in order to obtain a reliable system to be used for state prediction on rough terrains within the MPC optimization horizon.

State-space form of the vehicle model driven with traction force $u_p$ and steering momentum $u_s$

$$\dot{x} = f(x) + G(x) \begin{pmatrix} u_p \\ u_s \end{pmatrix} \tag{3.12}$$

can be written as

$$\begin{pmatrix} \dot{x}_{cg} \\ \dot{y}_{cg} \\ \dot{v} \\ \dot{\psi} \\ \ddot{\psi} \\ \dot{\beta} \\ \ddot{\varphi} \\ \dot{\delta} \end{pmatrix} = \begin{pmatrix} v f_{e_x} \\ v f_{e_y} \\ f_v \\ \dot{\psi} \\ f_\psi \\ f_\beta \\ f_\varphi \\ f_\delta \end{pmatrix} + \begin{pmatrix} 0 & 0 \\ 0 & 0 \\ g_{v_1} & 0 \\ 0 & 0 \\ 0 & 0 \\ g_{\beta_1} & 0 \\ 0 & 0 \\ 0 & g_{\delta_2} \end{pmatrix} \begin{pmatrix} u_p \\ u_s \end{pmatrix}, \tag{3.13}$$

where $\psi$ is the current vehicle orientation with respect to the world frame $xOy$, $\delta$ is the steering angle, $\varphi$ the body roll angle, and $\beta$ is the angle of the velocity direction with respect to the vehicle reference frame (Fig. 3.5).

The nonholonomic constraints are

$$f_{e_x} = \cos \beta \cos \psi - \sin \beta \sin \psi,$$
$$f_{e_y} = \cos \beta \sin \psi + \sin \beta \cos \psi.$$

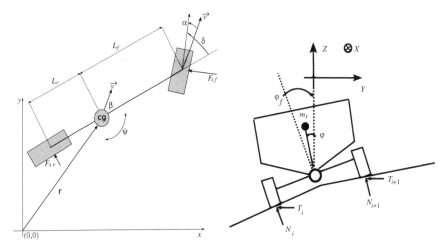

**Fig. 3.5** *Left* A schematic diagram of the vehicle model with slip angle. *Right* Suspension compliance model

$f_v$ and $g_{v_1}$ in (3.13) are obtained using the vehicle longitudinal dynamics motion equation (see Fig. 3.5left)

$$m\dot{v} = F_x \cos \beta + F_y \sin \beta, \qquad (3.14)$$

where $m$ is total vehicle mass, $F_x$ and $F_y$ are forces acting along the $x$ and $y$ directions of the vehicle internal reference frame, respectively, given by

$$F_x = -2 \sin \delta F_{s.f} + u_p, \qquad (3.15)$$

$$F_y = 2 \cos \delta F_{s.f} + 2F_{s.r} + \sum_{i=1}^{4} T_i, \qquad (3.16)$$

$F_{s.f}$ and $F_{s.r}$ being lateral forces of the front and rear wheels. These forces can be approximated for small slip angles using the stiffness coefficients, $C_f$ and $C_r$, as $F_f = C_f \alpha_f$ and $F_r = C_r \alpha_r$, $\alpha_f$ and $\alpha_r$ being the slip angles of the front and rear tyres approximated by

$$\alpha_f = \delta - \arctan \frac{v \sin \beta + L_f \dot{\psi}}{v \cos \beta}, \qquad (3.17)$$

$$\alpha_r = - \arctan \frac{v \sin \beta - L_r \dot{\psi}}{v \cos \beta}, \qquad (3.18)$$

where $L_f$ and $L_r$ are the distances of the front and rear wheels from the vehicle center of mass, respectively.

$T_i$ represents the terrain disturbance force acting at each wheel $i = 1..4$. Assuming the terrain elevation is a continuous and differentiable function $z(x, y)$, $T_i$ is given by

$$T_i = N_i \left( -\sin\psi \frac{\partial z}{\partial x_0} + \cos\psi \frac{\partial z}{\partial y_0} \right), \tag{3.19}$$

where $N_i$ is the normal contact force at wheel $i$ while $\frac{\partial z}{\partial x_0}$ and $\frac{\partial z}{\partial y_0}$ are gradients calculated in the vehicle body frame.

$f_\beta$ and $g_{\beta_1}$ are obtained using the momentum equation (see Fig. 3.5)

$$mv(\dot{\beta} + \dot{\psi}) = -F_x \sin\beta + F_y \cos\beta + m_s h \ddot{\varphi}, \tag{3.20}$$

$m_s$ being the mass of the chassis and $h$ the height of the chassis center of mass (see Fig. 3.5right).

$\ddot{\varphi}$ can be extracted from the suspension compliance model equation

$$I_{xx}\ddot{\varphi} = F_y h + M_{\text{roll}} + M_s, \tag{3.21}$$

where $I_{xx}$ is the roll moment inertia of the chassis, $M_{\text{roll}} = m_s g h \varphi$ is the moment caused by the inclination of the chassis center of mass (see the right Fig. 3.5), and $M_s$ the suspension moment on sloped terrain which can be given as

$$M_s = -k_f(\varphi - \varphi_f) - k_r(\varphi - \varphi_r) - b_f(\dot{\varphi} - \dot{\varphi}_f) - b_r(\dot{\varphi} - \dot{\varphi}_r), \tag{3.22}$$

$k_f$, $k_r$ being the stiffness and $b_f$, $b_r$ the damping rates of the respective axles. $\varphi_f$ and $\varphi_r$ are the roll disturbances caused by the terrain. By the assumption that wheels do not loose the contact with the terrain, these disturbances are given by

$$\varphi_{f,r} = \frac{z_{f,r+1} - z_{f,r}}{y_w}, \tag{3.23}$$

$$\dot{z}_{f,r} = \frac{\partial z}{\partial x_0} v \cos(\psi + \beta) + \frac{\partial z}{\partial y_0} v \sin(\psi + \beta) \tag{3.24}$$

$z_{f,r}$ are the positions of front and rear wheels respectively, and $y_w$ is the vehicle width.

$f_\psi$ is obtained using the momentum equation (see Fig. 3.5)

$$I_{zz}\ddot{\psi} = 2F_{s.f}L_f \cos\delta - 2F_{s.r}L_r + \sum_{i=1}^{4} T_i L_i, \tag{3.25}$$

$I_{zz}$ being the yaw moment of inertia, $L_i$ the longitudinal position of each wheel with respect to the vehicle center of mass.

The last row of (3.13) represents the steering dynamics of the vehicle from which $f_\delta$ and $g_{\delta_2}$ can be extracted.

## 3.3.2 Energy-Shaping Using a Navigation Function

The main task of this subsection is to obtain a virtual model from (3.13) which is passive with a globally asymptotically stable goal position. More precisely, it is required that the equilibrium point is $(x_{cg} \ y_{cg} \ v)_e = (x^* \ y^* \ 0)$, where $(x^*, y^*)$ is the goal position and $(\cdot)_e$ denotes an equilibrium point. The position coordinate transformation $e_x = x_{cg} - x^*$, $e_y = y_{cg} - y^*$, transforms the desired equilibrium point into the zero-state $(e_x \ e_y \ v)_e = (0 \ 0 \ 0)$.

Now, the model of the system given in (3.13) becomes

$$\dot{e} = f(e) + g(e) \begin{pmatrix} u_p \\ u_s \end{pmatrix}, \tag{3.26}$$

where

$$\dot{e} = (\dot{e}_x \ \dot{e}_y \ \dot{v} \ \dot{\psi} \ \ddot{\psi} \ \dot{\beta} \ \ddot{\varphi} \ \dot{\delta})^T, \ f(e) = f(x), \ g(e) = g(x).$$

The energy of the given system can be shaped with the following traction force input

$$u_p = \frac{1}{g_{v_1}} (-f_v - k\nabla NF(\mathbf{r})\mathbf{e}_{\dot{r}}) + v_p, \tag{3.27}$$

where

$$k\nabla NF(\mathbf{r})\mathbf{e}_{\dot{r}} = k\|\nabla NF(\mathbf{r})\| \cos \angle(\nabla NF(\mathbf{r}), \mathbf{e}_{\dot{r}}) \tag{3.28}$$

is the scaled inner product of the gradient of the navigation function $NF(\mathbf{r})$ with a unit vector of the current vehicle direction represented by the vector $\mathbf{e}_{\dot{r}}$. This term favors those directions that move the vehicle along the paths which decrease the value of the navigation function $NF(\mathbf{r})$, hence toward the goal position. For instance, if the vehicle goes in the direction of the steepest descent of the navigation function, (3.28) will have a minimum possible value thus providing the maximum value in (3.27). If the vehicle moves increasing the value of the navigation function, the component (3.28) will have positive values, hence decreasing the speed and stopping the vehicle. $v_p$ is the new control input of the system replacing the traction force.

In [5] the construction of the navigation function is discussed for the case when the vehicle only has information about its sensor range environment.

The virtual model obtained by the energy-shaping technique applied to model (3.26) is

$$\dot{e} = \tilde{f}(e) + \tilde{g}(e) \begin{pmatrix} v_p \\ u_s \end{pmatrix}, \tag{3.29}$$

where $\tilde{g}(e) = g(e)$ and

$$\tilde{f}(e) = (vf_{e_x} \ vf_{e_y} \ -k\nabla NF(\mathbf{r})\mathbf{e}_{\dot{r}} \ \dot{\psi} \ f_\psi \ \tilde{f}_\beta \ f_\varphi \ f_\delta)^T.$$

Note that

$$\dot{v} = -k\nabla \mathrm{NF}(\mathbf{r})\mathbf{e}_{\hat{r}} f_r + g_{v_1} v_p \tag{3.30}$$

and the function $f_\beta$ was changed into $\tilde{f}_\beta$ since $g_{\beta_1} \neq 0$.

In Chap. 2 [9] it has been shown that for the general case of the energy shaped vehicle model, the subsystem that contains the triple state of interest $x_{ss} = (e_x \ e_y \ v)^T$ has the zero-state equilibrium point, that is $x_{ss_e} = (0\ 0\ 0)^T$. For the purpose of clarity, the proof is recalled here.

Assuming $\dot{e}_x = 0$, $\dot{e}_y = 0$ and $\dot{v} = 0$, namely that there is no movement in both directions of the reference frame, the vehicle velocity is equal to zero, $v = 0$. From $\dot{v} = 0$, using (3.30) with input $v_p \equiv 0$, it follows $\nabla \mathrm{NF}(\mathbf{r})\mathbf{e}_{\hat{r}} = 0$. One possible solution of the latter equation, $\cos(\nabla \mathrm{NF}(\mathbf{r})\mathbf{e}_{\hat{r}})) = 0$, implies that the first condition in (2.8) is not satisfied since this equality also holds at the end of the operating time horizon $T_1$. This means that this equality is true only for the second possible solution, $\mathbf{r} = (x^* \ y^*)$, that is $e_x = 0$ and $e_y = 0$, since the unique minimum of the navigation function $\mathrm{NF}(\mathbf{r})$ is at this point.

### 3.3.3 Passivity-Based Stability

#### 3.3.3.1 Energy Storage Function

It was shown in Chap. 2 [9] that the energy storage function selected according to (2.2) satisfies the requirements given by Theorem 1 for the general vehicle model. For the new virtual system obtained by the energy-shaping technique, this means that the energy storage function $V(x)$ is radially unbounded and positive definite assuming the navigation function can be constructed such that it is continuously differentiable. For the case of the virtual model of the vehicle acting in rough terrain, condition (1.5) of Theorem 1 is satisfied

$$\begin{aligned}
\dot{V} &= \frac{\partial V}{\partial x} f(x) = [\frac{\partial V}{\partial \mathbf{e}_x} \ \frac{\partial V}{\partial \mathbf{e}_y} \ \frac{\partial V}{\partial v} \ \ \mathbf{0}] f(x) \\
&= [k\nabla \mathrm{NF}(\mathbf{r}) \ v \ \mathbf{0}] f(x) \\
&= k\nabla \mathrm{NF}(\mathbf{r})\dot{\mathbf{r}} + v(-k\nabla \mathrm{NF}(\mathbf{r})\mathbf{e}_{\hat{r}} f_r) = 0,
\end{aligned} \tag{3.31}$$

where

$$f(x) = (\dot{\mathbf{r}} \ \ -k\nabla \mathrm{NF}(\mathbf{r})\mathbf{e}_{\hat{r}} \ \ \dot{\psi} \ \ f_\psi \ \ \tilde{f}_\beta \ \ f_\varphi \ \ f_\delta)^T.$$

#### 3.3.3.2 System Output

In order to have a passive system, the output of the system is selected according to (2.4) and is given by

$$\mathbf{y}^T = \frac{\partial V}{\partial x} g(x) = \left[ \frac{\partial V}{\partial \mathbf{e}_x} \ \frac{\partial V}{\partial \mathbf{e}_y} \ \frac{\partial V}{\partial v} \ \mathbf{0} \right] G(x) \tag{3.32}$$

$$= [k\nabla NF(\mathbf{r}) \ v \ \mathbf{0}]G(x) = [g_{v_1} v \ 0]$$

This means that the output of interest for the given storage function is $y_1$

$$y_1 = g_{v_1} v = \frac{1}{m} v, \tag{3.33}$$

that will be used to define a damping feedback to the model through input $u_p$ only.

### 3.3.3.3 Zero-State Observability

In Chap. 2 [9] it has been shown that for the general case of the vehicle with energy-shaped model, the subsystem that contains the triple state of interest $x_{ss}$ was zero-state observable. For the purpose of clarity, this proof discussion is recalled here.

The subsystem of interest is

$$\begin{pmatrix} \dot{e}_x \\ \dot{e}_y \\ \dot{v} \end{pmatrix} = \begin{pmatrix} vf_{e_x} \\ vf_{e_y} \\ -k\nabla NF(\mathbf{r})\mathbf{e}_{\dot{r}} \end{pmatrix} + \begin{pmatrix} 0 & 0 \\ 0 & 0 \\ g_{v1} & 0 \end{pmatrix} \begin{pmatrix} v_p \\ u_s \end{pmatrix} \tag{3.34}$$

ZSO conditions $y \equiv 0$ and $u \equiv 0$ imply $v \equiv 0$ if $y_1$ is selected according to (3.33). This means $\dot{e}_x = 0$ and $\dot{e}_y = 0$, as well as $\dot{v} = 0$. Using (3.30), it follows $\nabla NF(\mathbf{r})\mathbf{e}_{\dot{r}} = 0$. Similar to the discussion given in Subsection B, the latter equation implies $\mathbf{r} = (x^* \ y^*)$, that is $e_x = 0$ and $e_y = 0$, so that $x_{ss} = 0$, namely the subsystem (3.34) is zero-state observable.

### 3.3.3.4 Damping Injection

Since all conditions of Theorem 1 are satisfied, the new traction force input $v_p$ can be selected in the form $v_p = -\phi(y)$, where $\phi$ is any locally Lipschitz function such that $\phi(0) = 0$ and $y^T\phi(y) > 0$ for all $y \neq 0$. One possible choice of damping injection using function $\phi(y)$ is (see e.g., [7])

$$v_p = -\epsilon \frac{1}{g_{v_1}} \frac{2}{\pi} \arctan(k_v v) = -\epsilon m \frac{2}{\pi} \arctan(k_v v), \tag{3.35}$$

where $\epsilon$ and $k_v$ are positive constants to be selected.

In order to obtain stability, by the assumption $v \geq 0$, one can write

$$v_p \leq -\epsilon m \frac{2}{\pi} \arctan(k_v v). \tag{3.36}$$

This choice satisfies (2.5) making the equilibrium $x_{ss_e}$ globally asymptotically stable.

This claim can be easily verified finding the time derivative of the energy storage function $V$ along the trajectories of system (3.29) under condition (3.36). In fact, using (3.30)

$$\dot{V} = \frac{\partial V}{\partial x}\dot{x} = [\frac{\partial V}{\partial \mathbf{r}} \quad \frac{\partial V}{\partial v} \quad \mathbf{0}]\dot{x} = [k\nabla NF(\mathbf{r}) \quad v \quad \mathbf{0}]\dot{x}$$
$$= k\nabla NF(\mathbf{r})\dot{\mathbf{r}} + v(-k\nabla NF(\mathbf{r})\mathbf{e}_{\dot{r}}\,f_r + g_{v_1}v_p) = g_{v_1}vv_p,$$

and, under constraint (3.36), the condition on the derivative of the energy storage function along the trajectories of the closed loop system is obtained as follows

$$\dot{V} = g_{v_1}vv_p \leq -\epsilon v \frac{2}{\pi} \arctan(k_v v). \tag{3.37}$$

Hence, $\dot{V}$ is negative semidefinite and $\dot{V} = 0$ if and only if $v = 0$. By zero-state observability, $y \equiv 0$ and $u \equiv 0$ implies $x_{ss} = 0$. Therefore, by the invariance principle, the origin of the subsystem represented by $x_{ss}$ is globally asymptotically stable. This means that the energy of the system given by the virtual potential $kNF(\mathbf{r})$ and the kinetic energy term $\frac{1}{2}v^2$ will decrease along the trajectories of the system (3.29), moving the vehicle toward the goal position.

### 3.3.4 Simulation

The vehicle model which includes terrain effects along the MPC time horizon is assumed to be known in order to appropriately shape the energy of the vehicle to

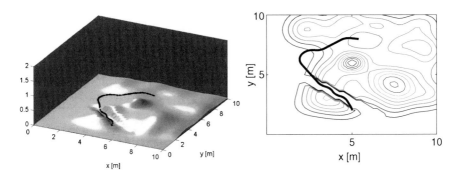

**Fig. 3.6** Vehicle follows flat terrain on the left-side toward the goal position

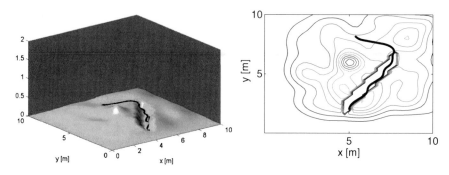

**Fig. 3.7** Vehicle follows flat terrain section on the right-side toward the goal position

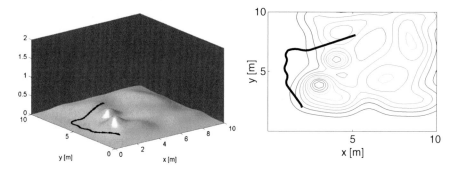

**Fig. 3.8** Vehicle avoids rough terrain sections toward the goal position

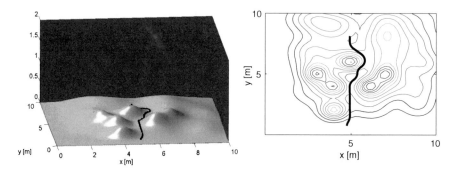

**Fig. 3.9** Vehicle avoids highly rough terrain sections toward the goal position

provide feasible trajectories. Uncertainties caused by sensor measurements of the terrain effects may be compensated using a path tracking controller. Figures 3.6, 3.7, 3.8, 3.9 and 3.10 illustrate the vehicle capability to avoid difficult terrain sections while approaching the goal position assuming completely known terrain. All presented examples are illustrated by two subfigures. The left one denotes the generated path, while the right one depicts the contour plot representing the cost field of the level of roughness. Figures 3.6 and 3.7 show the capability of the vehicle to follow

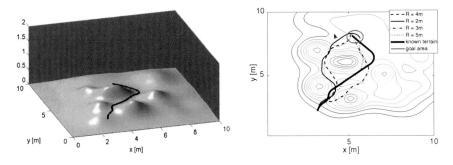

**Fig. 3.10** *Left* Vehicle avoids rough terrain sections in known terrain. *Right* Vehicle avoids rough terrain sections using different limited sensor ranges, R

flat parts of the terrain toward the goal position. In Fig. 3.9, a different rough terrain is given where the vehicle avoids more difficult areas while moving toward the goal position.

For the case when the terrain roughness is unknown outside the sensor range, the unknown area is considered to be completely flat when the roughness-to-go value is estimated. Some different cases are shown in Fig. 3.10(right). An improvement in estimation of the roughness-to-go value at the end of the optimization horizon has been presented in [10, 11]. For the case of obstacle-free terrain but of unknown roughness, any quadratic function can be used to form the navigation function having a unique minimum at the goal position. However, for the terrain with obstacles, any form of the navigation function can be used instead [6].

# References

1. P. Oegren, N.E. Leonard, A convergent dynamic window approach to obstacle avoidance. IEEE Trans. Robot. **21**(2), 188–195 (2005)
2. E. Bakker, L. Nyborg, H.B. Pacejka, *Tyre modeling for use in vehicle dynamics studies* (Society of Automotive Engineers, Warrendale, 1987)
3. H.B. Pacejka, *Tire and Vehicle Dynamics*, 2nd edn. (Society of Automotive Engineers, Warrendale, 2006)
4. Y. Yoon, J. Shin, H.J. Kim, Y. Park, S. Sastry, Model-predictive active steering and obstacle avoidance for autonomous ground vehicles. Control Eng. Pract. **17**(7), 741–750 (2009)
5. O. Brock, O. Khatib, High-speed navigation using the global dynamic window approach. Proc. IEEE Int. Conf. Robot. Autom. **1**, 341–346 (1999)
6. J. Latombe, *Robot Motion Planning* (Kluwer, Boston, 1991)
7. H.K. Khalil, *Nonlinear Systems*, 3rd edn. (Prentice Hall, Upper Saddle River, 2002)
8. S. Peters, K. Iagnemma, Mobile robot path tracking of aggressive maneuvers on sloped terrain, in *Proceedings of the IEEE/RSJ International Conference on Intelligent Robots and Systems*, Washington, 2008
9. A. Tahirovic, G. Magnani, P. Rocco, Mobile robot navigation using passivity-based MPC, in *Proceedings of the IEEE/ASME International Conference on Advanced Intelligent Mechatronics*, pp. 248–488, 2010

10.  A. Tahirovic, G. Magnani, A roughness-based RRT for mobile robot navigation planning, in
     *Proceedings of the 18th IFAC World Congress*, pp. 5944–5949, 2011
11.  A. Tahirovic, G. Magnani, Y. Kuwata, An approximate of the cost-to-go map on rough terrains,
     in *Proceedings of the IEEE International Conference on Mechatronics*, 2013

# Chapter 4
# Some Limitations and Real-Time Implementation

## 4.1 Introduction

This chapter gives an analysis of the worst possible case which the vehicle might experience during the task execution on rough terrains while using the PB/MPC motion planner. Additionally, we present a possible real-time implementation of an MPC-like motion planner using algorithms developed for optimal control problems.

## 4.2 The Worst Case Scenarios on Rough Terrains

In the following analysis it is assumed that $k_v$ has taken large providing the term $\frac{2}{\pi}\arctan(k_v v)$ be approximately a unit Heaviside function for $v \geq 0$. Therefore, (3.37) can be approximately rewritten as

$$\dot{V} \leq -\epsilon v. \tag{4.1}$$

### 4.2.1 Unknown Rough Terrain with Obstacles

The following theorem gives the maximum time needed by the vehicle to reach the goal position. It is assumed that after new terrain information appears and the new navigation function is constructed, there exists a feasible solution to the PB/MPC navigation scheme for rough terrains. If this is not the case, the vehicle should perform a turning maneuver toward the decrease of the navigation function.

If $v_{av}$ is the average velocity along the whole path, $v_{min} \neq 0$ is the smallest feasible velocity of the moving vehicle, $k$ and $\epsilon$ are parameters of the PB/MPC optimization framework, $M$ the number of different recalculations of the navigation functions caused by the change in perception of the terrain, and $\Delta NF^j$ the decrease of the

A. Tahirovic and G. Magnani, *Passivity-Based Model Predictive Control for Mobile Vehicle Motion Planning*, SpringerBriefs in Control, Automation and Robotics, DOI: 10.1007/978-1-4471-5049-7_4, © The Author(s) 2013

navigation function from two successive recalculations, then the theorem is stated as follows.

**Theorem 4** (*Time and Path Length*) *If the PB/MPC scheme is used for the navigation of mobile vehicles in unknown rough terrain with obstacles assuming zero initial velocity, then the longest path, the final time $T_{goal}$, and the upper time bound $T_{goal-max}$ needed for the task completion are given as follows*

$$l \leq l_{max} = \frac{k}{\epsilon} \sum_{j=1}^{M} \Delta NF^j, \qquad (4.2)$$

$$T_{goal} = \frac{k}{\epsilon v_{av}} \sum_{j=1}^{M} \Delta NF^j, \qquad (4.3)$$

$$T_{goal-max} = \frac{k}{\epsilon v_{min}} \sum_{j=1}^{M} \Delta NF^j, \qquad (4.4)$$

*where $T_{goal} \leq T_{goal-max}$.*

*Proof* Integration of both sides of the condition (4.1) along the $i$th time horizon, $t \in (t_{i-1}, t_{i-1} + T_1 = t_i)$, and within $j$th navigation function, $NF^j$, yields $V^j(t_i) - V^j(t_{i-1}) \leq -\epsilon l_i^j$, where $l_i^j$ is the arc length of the traversed path segment within the time horizon. By taking the expression for the energy storage function given by (2.2), the following inequality can be obtained

$$k(NF^j(\mathbf{r}_{i-1}) - NF^j(\mathbf{r}_i)) + \frac{1}{2}(v_{i-1}^2 - v_i^2) \geq \epsilon l_i^j$$

$$\Leftrightarrow k\delta NF_i^j + \frac{1}{2}(v_{i-1}^2 - v_i^2) \geq \epsilon l_i^j, \qquad (4.5)$$

where $r_{i-1}$ and $r_i$ are the positions of the vehicle at the beginning and the end of the $i$th horizon, while $\delta NF^j = NF^j(r_{i-1}) - NF^j(r_i)$ is the decrease of the navigation function along the $i$th horizon when $NF^j$ represents the terrain.

By performing the sum of both the left and right sides of (4.5) along the whole path from the initial to the final position of the current active navigation function $NF^j$, that is the position where the next navigation function is calculated, it follows

$$\sum_{i=1}^{N} \left[ k\delta NF_i^j + \frac{1}{2}(v_{i-1}^2 - v_i^2) \right] \geq \sum_{i=1}^{N} \epsilon l_i^j$$

$$\Leftrightarrow k\Delta NF^j + \frac{1}{2}\left(v_0^{2(j)} - v_{final}^{2(j)}\right) \geq \epsilon l^j, \qquad (4.6)$$

where $N$ is the number of the performed optimization horizons within $\text{NF}^j$, $l^j$ is the length of the selected path from the initial to the final position within $\text{NF}^j$, $\Delta\text{NF}^j$ is the whole decrease of the navigation function $\text{NF}^j$, and $v_0^j$, $v_{\text{final}}^j$ the initial and the final velocity of the selected path within $\text{NF}^j$.

If both sides of (4.6) are summed along the whole path from the initial to the goal position within each navigation function $\text{NF}^j$, it results

$$\sum_{j=1}^{M}\left[k\Delta\text{NF}^j + \frac{1}{2}\left(v_0^{2(j)} - v_{\text{final}}^{2(j)}\right)\right] \geq \sum_{j=1}^{M}\epsilon l^j$$

$$\Leftrightarrow k\sum_{j=1}^{M}\Delta\text{NF}^j + \frac{1}{2}\left(v_0^{2(1)} - v_{\text{final}}^{2(M)}\right) \geq \epsilon l. \qquad (4.7)$$

Since the vehicle started with zero velocity $v_0^1 = 0$ as it was assumed in Theorem 4, and the velocity at the goal position $v_{\text{final}}^M = 0$ according to the PB/MPC navigation scheme, the latter inequality gives the maximum length of the PB/MPC generated path (4.2).

By taking $l = T_{\text{goal}}v_{\text{av}}$ and $l_{\text{max}} = T_{\text{goal}}v_{\text{min}}$, equations (4.3) and (4.4) easily follow.

### 4.2.2 Completely Known Rough Terrain with Obstacles

If the terrain is completely known then the navigation function is computed only once, hence $M = 1$.

**Corollary** (*Time and Path Length*) *If the PB/MPC navigation scheme is used for the navigation of mobile vehicle in completely known rough terrain with obstacles assuming zero initial velocity, then the longest path, the final time $T_{\text{goal}}$, and the upper time bound $T_{\text{goal}-\text{max}}$ needed for the task completion are given as follows*

$$l \leq l_{\text{max}} = \frac{k}{\epsilon}\text{NF}(\mathbf{r}_0), \qquad (4.8)$$

$$T_{\text{goal}} = \frac{k}{\epsilon v_{\text{av}}}\text{NF}(\mathbf{r}_0), \qquad (4.9)$$

$$T_{\text{goal}-\text{max}} = \frac{k}{\epsilon v_{\text{min}}}\text{NF}(\mathbf{r}_0), \qquad (4.10)$$

*where $T_{\text{goal}} \leq T_{\text{goal}-\text{max}}$.*

*Proof* The proof directly follows from (4.2–4.4) since $M = 1$ and

$$\text{NF}^1 - \text{NF}^0 = \text{NF}(r_0) - \text{NF}(r_{\text{goal}}) = \text{NF}(r_0).$$

If $\text{NF} = \text{NF}(x, y)$ is the navigation function given by Cartesian coordinates, and $\text{NF} = \text{NF}(\rho, \theta)$ being the navigation function given by polar coordinates, where $x = \rho \cos \theta$ and $y = \rho \sin \theta$, then the following theorem gives the shape of the worst case scenario generated by the PB/MPC optimization framework for completely known rough terrain with obstacles.

**Theorem 5** (*Path Shape*) *If the PB/MPC scheme is used for the navigation of a mobile vehicle in completely known rough terrain with obstacles, assuming zero initial velocity, the final path shape of the worst case scenario can be generated by the differential equations (4.11) and (4.12)*

$$\frac{\partial \text{NF}}{\partial x} + \frac{\partial \text{NF}}{\partial y}\frac{dy}{dx} = -\frac{\epsilon}{k}\sqrt{1 + \left(\frac{dy}{dx}\right)^2} \tag{4.11}$$

$$\frac{\partial \text{NF}}{\partial \rho}\frac{d\rho}{d\theta} + \frac{\partial \text{NF}}{\partial \theta} = -\frac{\epsilon}{k}\sqrt{\frac{d\rho}{d\theta} + \rho^2}. \tag{4.12}$$

*Proof* In order to obtain the shape of the worst case scenario for which the vehicle moves with constant smallest feasible velocity $v_{\min}$, one can start from (4.1) and obtain

$$dV = kd\text{NF}(\mathbf{r}) + v dv \le -\epsilon dr. \tag{4.13}$$

Since we assume $v = v_{\min}$ along the path, that is $dv = 0$, (4.13) can be rewritten as

$$d\text{NF}(\mathbf{r}) \le -\frac{\epsilon}{k}dr. \tag{4.14}$$

For the worst case scenario the infinitesimal decrease $d\text{NF}$ of the navigation function will be equal to the right hand side.

In the case when $\text{NF} = \text{NF}(x, y)$, the differential equation (4.11) from which the shape of the longest possible path can be directly generated follows from (4.14) using the total differential of the function NF, namely

$$\frac{\partial \text{NF}}{\partial x}dx + \frac{\partial \text{NF}}{\partial y}dy = -\frac{\epsilon}{k}\sqrt{(dx)^2 + (dy)^2}. \tag{4.15}$$

Similarly, if $\text{NF} = \text{NF}(\rho, \theta)$, where $dr = \sqrt{\frac{d\rho}{d\theta} + \rho^2}d\theta$, then the differential equation (4.12), which is equivalent to the previous one, can be obtained

$$\frac{\partial \text{NF}}{\partial \rho}d\rho + \frac{\partial \text{NF}}{\partial \theta}d\theta = -\frac{\epsilon}{k}\sqrt{\frac{d\rho}{d\theta} + \rho^2}d\theta. \tag{4.16}$$

Note that the coefficients of equations (4.11) and (4.12) depend on the terrain configuration.

### 4.2.3 Unknown Rough Terrain Without Obstacles

Quite interesting is the case of obstacle-free terrain but with areas of different level of roughness. Here, any quadratic function can be used to form the navigation function having a unique minimum at the goal position.

One particular example is when the navigation function represents the distance from the current position to the goal position, that is $NF(\rho) = \rho$.

**Corollary** (*Path Shape and Length*) *If the PB/MPC navigation scheme is used for the navigation of a mobile vehicle in unknown rough terrains without obstacles assuming the navigation function represents the shortest distance to the goal position and zero initial velocity, the final path shape of the worst case scenario can be generated by the differential equations (4.17)*

$$\theta = \theta_0 - \frac{1}{p}\left[\ln|t| + \frac{2p}{t+p} - \ln|t_{\rho_0}| - \frac{2p}{t_{\rho_0}+p}\right], \tag{4.17}$$

*where* $p = \frac{\epsilon}{k}$, $t = -2\rho - \sqrt{p^2 + 4\rho^2}$, *and* $\theta_0$, $t_{\rho_0} = t(\rho_0)$, $NF(\mathbf{r}_0) = \rho_0$ *being the initial values given the starting position of the vehicle, and the bounds of the path length being*

$$\rho_0 \le l \le \frac{k}{\epsilon}\rho_0. \tag{4.18}$$

*Proof* In this case, the differential equation (4.12) becomes

$$\frac{d\rho}{d\theta} = -\frac{\epsilon}{k}\sqrt{\frac{d\rho}{d\theta} + \rho^2}. \tag{4.19}$$

The explicit solution to this differential equation can be easily found in the form Eq. (4.17) as follows. Eq. (4.19) implies the following equation

$$y^2 - p^2 y - p^2 \rho = 0, \tag{4.20}$$

where $y = \frac{d\rho}{d\theta} \le 0$ and $p = \frac{\epsilon}{k}$. The solution of this equation can be expressed as

$$y = \frac{p^2 - p\sqrt{p^2 + 4\rho^2}}{2}, \tag{4.21}$$

from which follows

$$\frac{d\rho}{p - \sqrt{p^2 + 4\rho^2}} = \frac{p}{2} d\theta. \tag{4.22}$$

If both left and right sides of the latter equation are integrated over the interval $(\rho_0, \rho)$ and $(\theta_0, \theta)$, respectively, the right integral is easily solved

$$\int_{\theta_0}^{\theta} \frac{p}{2} d\theta = \frac{p}{2}(\theta - \theta_0), \tag{4.23}$$

while the left one is the integral of irrational function,

$$\int_{\rho_0}^{\rho} \frac{d\rho}{p - \sqrt{p^2 + 4\rho^2}}, \tag{4.24}$$

which can be solved by the substitution

$$p - \sqrt{p^2 + 4\rho^2} = 2\rho + t + p. \tag{4.25}$$

From this substitution one can easily obtain

$$\rho = \frac{p^2 - t^2}{4t}, \tag{4.26}$$

$$t = -2\rho - \sqrt{p^2 + 4\rho^2}, \tag{4.27}$$

and

$$d\rho = -\frac{t^2 + p^2}{4t^2} dt. \tag{4.28}$$

Finally, the integral (4.24) becomes the integral of the rational function

$$-\frac{1}{2} \int_{t_0}^{t} \frac{t^2 + p^2}{t(t + p)^2}, \tag{4.29}$$

whose $t_0$ and $t$ are defined by the substitution (4.27). This integral can be easily solved and its solution is given by

$$\int_{\rho_0}^{\rho} \frac{d\rho}{p - \sqrt{p^2 + 4\rho^2}} = -\frac{1}{2}\left( \ln|t| - \ln|t_0| + \frac{2p}{t + p} - \frac{2p}{t_0 + p} \right). \tag{4.30}$$

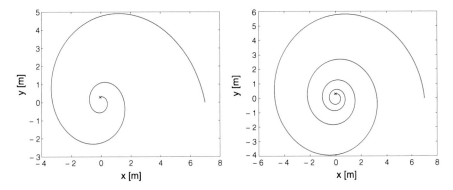

**Fig. 4.1** Shape of the path generated for the worst case scenario of the PB/MPC framework when there are no obstacles. *Left $\epsilon = 0.2$, Right $\epsilon = 0.1$*

Finally, the solution (4.17) of the differential equation (4.19) follows from the equality of the solutions (4.23) and (4.30), and the substitution (4.27).

As the shortest path from the vehicle initial position $\mathbf{r}_0$ to the goal position is $\rho_0$, (4.18) directly follows from (4.8).

As one can expect this shape has the spiral form. Diagrams for different $\epsilon$, which is a parameter to be selected in the PB/MPC framework, are given in Fig. 4.1. Note that there is a tradeoff between the roughness sensitivity of the algorithm and the length of the generated path which can be adjusted by parameter $\epsilon$. This was expected since the solution was derived from (4.1), wherein the larger $\epsilon$ the faster the energy storage function decreases.

## 4.3 Real Time Implementation of an MPC Based Motion Planner

Finding an optimal path on rough terrains, given a vehicle model and all information about the terrain, can be expressed as a two point boundary value optimal control problem (OCP). Including the terrain shape into an objective function for the OCP might result into a problem difficult to solve. Namely, the OCP softwares, including ACADO [1], the software used in this work, require a differentiable objective function. To overcome this problem, a kind of interpolation of the terrain shape must be applied. However, such an interpolation might be computationally intensive even for medium size terrains, and finding the best path solving an OCP might be impractical for real-time implementation.

The MPC optimization problem can be expressed as an initial value OCP problem with an end-free position (2.1–2.9). For the purpose of the motion planning on rough terrains, we locally interpolate roughness data in order to construct differentiable objective functions (2.1) for each optimization cycle required by the ACADO. Local interpolation is being conducted for local roughness measure, $\gamma(\mathbf{x}, \mathbf{u})$, as well for the cost-to-go term, $\Gamma(t_0 + T)$. The differentiable function for the local roughness measure is constructed from the roughness map provided by the vehicle sensors, while the differentiable cost-to-go term is interpolated from the numerical map which can be obtained by the Dijkstra's algorithm or in an approximated fashion as in [2, 3]. The latter is especially important in case of large-scale terrains [3], where computing an optimal value function for the whole terrain is rather computationally complex task. The local construction of a differentiable objective function significantly decreases the computational complexity of any interpolation technique in accordance to a requirement for having a differentiable function over the whole terrain as in case of a two point boundary problem. We have used the interpolation technique for terrain shapes presented in [4]. In addition, the numerical map can be updated during the task execution due to the new environment information as in [5].

In some rare cases for which ACADO fails, bringing back an infeasible solution due to the MPC constraints, we use a backup strategy to guide the vehicle forward. In those cases, a planner selects a close way-point which is located along the steepest descent of the cost-to-go term, $\Gamma(t_0 + T)$, and solves for a two point boundary value OCP problem.

In the sequel, the advantages of an MPC motion planner implementation are summarized. An MPC-based motion planner can easily accommodate for a vehicle model with all the required constraints. The planner might be near optimal (giving the current state information) due to "the optimality principle" if $\Gamma(t_0 + T)$ is a near optimal estimator of the cost-to-go optimization term. Since the MPC horizon can be arbitrarily chosen, a terrain shape interpolation required to get a differentiable objective function can be locally applied using algorithms presented in [4]. Having a differentiable objective function allows for using an OCP software. Using a software to solve a local OCP problem, like ACADO, covers much of the control and state space comparing to [6–10]. Finally, instead of repeating the complete path planning procedure from scratch when the vehicle senses new information, the cost-to-go term, $\Gamma(t_0 + T)$, can be easily updated similarly to [5].

### 4.3.1 Simulation Results

The example presented in Fig. 4.2 shows that the generated path avoids obstacles, follows less roughness regions (blue regions), and reaches the goal position (start and goal positions are marked with a red and a pink disk, respectively).

In some cases where the terrain is small scale, it is even possible to compute an optimal solution (solving two point boundary problem) in a reasonable time by an OCP software such as ACADO. For this reason, we have used a small terrain $50 \times 50$ m to compare an optimal and an MPC-based solutions exploring the MPC sub-optimality. Figure 4.3 depicts 10 simulations in which the same rough terrain and different vehicle initial positions are used. The average sub-optimality of the MPC-based path planner can be computed as

**Fig. 4.2**  An example of an
MPC-based solution

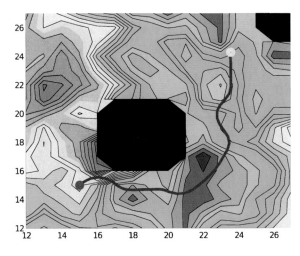

**Fig. 4.3**  1: Small-scale terrain. MPC and OCP solutions

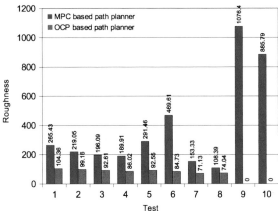

$$\alpha = \frac{1}{N} \sum \frac{\text{roughness}^{\text{OCP}}}{\text{roughness}^{\text{MPC}}} = 0.43$$

where $N$ is the number of simulations. One might see that in the 9th and 10th
simulations, ACADO did not find a feasible solution for the OCP problem (depicted
by 0 in the picture).

Figure 4.4 depicts another example with 10 simulations on the same terrain with
the same vehicle initial position and roughness shape, but with different obstacles.
There are some examples where an MPC-based solution has given a better result. This
can be explained by the fact that an OCP software parameterizes the control space in
order to find the best solution. This might produce a solution that is not necessary the
optimal one. In this example, the sub-optimality of the MPC path planner is much
higher ($\alpha = 0.93$).

**Fig. 4.4** II: Small-scale terrain. MPC and OCP solutions

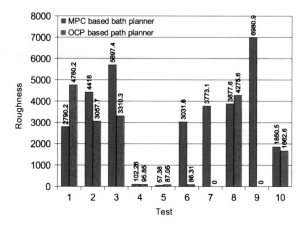

**Fig. 4.5** Large-scale terrain.
MPC and smooth gradient-
based solutions

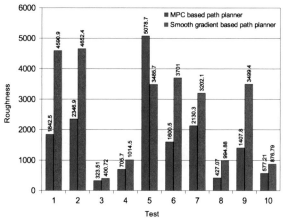

A two boundary value problem is difficult to solve in a feasible time on a large-scale terrain. For this reason, we use three different planners for a $500 \times 500$ m terrain, an MPC-based planner, a gradient-based planner, and a smooth gradient-based planner. The gradient-based planner is generated by the steepest descent of the cost-to-go function $\Gamma(t_0 + T)$. As already discussed, the gradient-based planner is not considered as an acceptable solution in our work, since it does not take the vehicle model into account, and it is hard to predict how well the vehicle will follow such path. However, in order to validate the MPC-based path planner, we introduce a smooth gradient-based path planner which picks a point on the path obtained by the gradient-based path planner and solves for a two boundary problem. Then, it repeats the procedure going toward the goal position. Figure 4.5 compares the two planners on 10 different rough terrains. The sub-optimality of the MPC-based path planner is $\alpha = 1.8$, which means that the MPC-based planner performs better than the smooth gradient path planner. Again, this can be explained by the fact that the smooth gradient-based path planner does take the vehicle model into account but only to follow the gradient-based path planner.

# References

1. B. Houska, H. Ferreau, ACADO toolkit—automatic control and dynamic optimization, http://acadotoolkit.org
2. A. Tahirovic, G. Magnani, A roughness-based RRT for mobile robot navigation planning. in *Proceedings of the 18th IFAC World Congress*, pp. 5944–5949 (2011)
3. A. Tahirovic, G. Magnani, Y. Kuwata, An approximate of the cost-to-go map on rough terrains. in *Proceedings of the IEEE International Conference on Mechatronics* (2013)
4. F.J. Aguilar, F. Agera, M.A. Aguilar, F. Carvajal, Effects of terrain morphology, sampling density, and interpolation methods on grid DEM accuracy. Photogram. Eng. Remote Sens. **71**(7), (2005), pp. 805–816
5. A. Stentz, The focussed D* algorithm for real-time replanning. in *Proceedings of the International Joint Conference on Artificial Intelligence*, pp. 1652–1659 (1995)
6. C.J. Green, A. Kelly, Toward optimal sampling in the space of paths. in *Proceedings of the International Symposium of Robotics Research*, pp. 171–180 (2007)
7. M. Pivtoraiko, R.A. Knepper, A. Kelly, Differentially constrained mobile robot motion planning in state lattices. J. Field Robot. **26**(3), 308–333 (2009)
8. T.M. Howard, A. Kelly, Optimal rough terrain trajectory generation for wheeled mobile robots. Int. J. Rob. Res. **26**(2), 141–166 (2007)
9. M. Pivtoraiko, A. Kelly, Efficient constrained path planning via search in state lattices. in *Proceedings of the 8th International Symposium on Artificial Intelligence, Robotics and Automation in Space*, vol. 8, Sept 2005
10. T.M. Howard, C.J. Green, A. Kelly, D. Ferguson, State space sampling of feasible motions for high-performance mobile robot navigation in complex environments. J. Field Robot. **25**(10), 325–345 (2008)

# Chapter 5
# Conclusion

The presented PB/MPC motion planning approach is based both on the energy shaping technique using a navigation function obtained from the terrain configuration and on the passivity-based MPC concept. The planner can be seen as a generalized DWA planning technique. The PB/MPC algorithm is a straightforward procedure that can be easily adapted to the navigation for a broad class of vehicles and terrains. It guarantees task completion under the assumption that the vehicle model is known and its states are obtainable through measurement and estimation at the end of each optimization interval.

The work presents the PB/MPC planner both for indoor and outdoor environments. The final form of the planner is obtained by the energy-shaping technique applied to the vehicle dynamics model using a navigation function, where the navigation function provides the information on the goal position and obstacles. The MPC is enhanced by the passivity constraint in order to stabilize the goal position guaranteeing the task completion, and it assures a safe driving policy which is consistent with the safe driving policy adopted by humans in such terrains.

The PB/MPC navigation and motion planner may easily use a truly complex vehicle and terrain model to generate feasible trajectories during the task execution. Having a possibility to use a more complex vehicle model during the planning stage of the vehicle navigation certainly provides a more reliable planner in comparison to the approaches that use a simplified model for trajectory generation. Consequently, the trajectories generated by the more complex navigation model are easier to track by the real vehicle. Such a feature inherently implies a more safe planning procedure in terms of collision-free paths.

An inherited property of the MPC optimization allows one to impose a wide range of additional constraints into the PB/MPC navigation. This property provides a possibility to further extend the PB/MPC framework including constraints on the vehicle stability that can be described by rollover and sideslip angles. The MPC cost function represents a cost value helping the vehicle to select smoother areas toward the goal position. An additional improvement of the cost-to-go estimation

A. Tahirovic and G. Magnani, *Passivity-Based Model Predictive Control for Mobile Vehicle Motion Planning*, SpringerBriefs in Control, Automation and Robotics, DOI: 10.1007/978-1-4471-5049-7_5, © The Author(s) 2013

might allow for further improvement of the planner in terms of mobility index. This extension will be the focus of the future work.

The planning PB/MPC framework is developed for a rather general vehicle model, it can also be used for other mobile systems such as aerial and underwater unmanned vehicles. As such, the PB/MPC algorithm can be seen as a broadly general planner suitable for a wide range of unmanned vehicles and environments.

# Editors' Biography

**Tamer Başar** is with the University of Illinois at Urbana-Champaign, where he holds the academic positions of Swanlund Endowed Chair, Center for Advanced Study, Professor of Electrical and Computer Engineering, Research Professor at the Coordinated Science Laboratory, and Research Professor at the Information Trust Institute. He received the B.S.E.E. degree from Robert College, Istanbul, and the M.S., M.Phil, and Ph.D. degrees from Yale University. He has published extensively in systems, control, communications, and dynamic games, and has current research interests that address fundamental issues in these areas along with applications such as formation in adversarial environments, network security, resilience in cyber-physical systems, and pricing in networks.

In addition to his editorial involvement with these *Briefs*, Başar is also the Editor-in-Chief of *Automatica*, Editor of two Birkhäuser Series on *Systems & Control* and *Static & Dynamic Game Theory,* the Managing Editor of the *Annals of the International Society of Dynamic Games* (ISDG), and member of editorial and advisory boards of several international journals in control, wireless networks, and applied mathematics. He has received several awards and recognitions over the years, among which are the Medal of Science of Turkey (1993); Bode Lecture Prize (2004) of IEEE CSS; Quazza Medal (2005) of IFAC; Bellman Control Heritage Award (2006) of AACC; and Isaacs Award (2010) of ISDG. He is a member of the US National Academy of Engineering, Fellow of IEEE and IFAC, Council Member of IFAC (2011-14), a past President of CSS, the Founding President of ISDG, and President of AACC (2010-11).

**Antonio Bicchi** is Professor of Automatic Control and Robotics at the University of Pisa. He graduated at the University of Bologna in 1988 and was a postdoc scholar at M.I.T. A.I. Lab between 1988 and 1990.

His main research interests are in:

- dynamics, kinematics, and control of complex mechanical systems, including robots, autonomous vehicles, and automotive systems;
- haptics and dextrous manipulation; and
- theory and control of nonlinear systems, in particular hybrid (logic/dynamic, symbol/signal) systems.

A. Tahirovic and G. Magnani, *Passivity-Based Model Predictive Control for Mobile Vehicle Motion Planning, SpringerBriefs in Control, Automation and Robotics,* DOI: 10.1007/978-1-4471-5049-7, © The Author(s) 2013

He has published more than 300 papers on international journals, books, and refereed conferences.

Professor Bicchi currently serves as the Director of the Interdepartmental Research Center "E. Piaggio" of the University of Pisa, and President of the Italian Association or Researchers in Automatic Control. He has served as Editor in Chief of the Conference Editorial Board for the IEEE Robotics and Automation Society (RAS), and as Vice President of IEEE RAS, Distinguished Lecturer, and Editor for several scientific journals including the *International Journal of Robotics Research*, the *IEEE Transactions on Robotics and Automation*, and *IEEE RAS Magazine*. He has organized and co-chaired the first World Haptics Conference (2005), and Hybrid Systems: Computation and Control (2007). He is the recipient of several best paper awards at various conferences, and of an Advanced Grant from the European Research Council. Antonio Bicchi has been an IEEE Fellow since 2005.

**Miroslav Krstic** holds the Daniel L. Alspach chair and is the Founding Director of the Cymer Center for Control Systems and Dynamics at University of California, San Diego. He is a recipient of the PECASE, NSF Career, and ONR Young Investigator Awards, as well as the Axelby and Schuck Paper Prizes. Professor Krstic was the first recipient of the UCSD Research Award in the area of engineering and has held the Russell Severance Springer Distinguished Visiting Professorship at UC Berkeley and the Harold W. Sorenson Distinguished Professorship at UCSD. He is a Fellow of IEEE and IFAC. Professor Krstic serves as Senior Editor for *Automatica* and *IEEE Transactions on Automatic Control* and as Editor for the Springer series *Communications and Control Engineering*. He has served as Vice President for Technical Activities of the IEEE Control Systems Society. Krstic has co-authored eight books on adaptive, nonlinear, and stochastic control, extremum seeking, control of PDE systems including turbulent flows, and control of delay systems.

Printed by Publishers' Graphics LLC
MLSI130426.15.16.80